LAGRANGIAN OPTICS

LAGRANGIAN OPTICS

by

Vasudevan Lakshminarayanan
University of Missouri-St. Louis
St. Louis Missouri

Ajoy K. Ghatak
Indian Institute of Technology
New Delhi, India

K. Thyagarajan
Indian Institute of Technology
New Delhi, India

SPRINGER SCIENCE+BUSINESS MEDIA, LLC

Library of Congress Cataloging-in-Publication Data

A C.I.P. Catalogue record for this book is available from the Library of Congress

ISBN 978-1-4613-5690-5 ISBN 978-1-4615-1711-5 (eBook)
DOI 10.1007/978-1-4615-1711-5

Printed on acid-free paper.

This book is dedicated to our wives:

Lorraine L Janeczko
Gopa Ghatak
Raji Thyagarajan

Contents

Preface

In geometrical optics, light propagation is analyzed in terms of light rays which define the path of propagation of light energy in the limit of the optical wavelength tending to zero. Many features of light propagation can be analyzed in terms of rays, of course, subtle effects near foci, caustics or turning points would need an analysis based on the wave nature of light.

All of geometric optics can be derived from Fermat's principle which is an extremum principle. The counterpart in classical mechanics is of course Hamilton's principle. There is a very close analogy between mechanics of particles and optics of light rays. Much insight (and useful results) can be obtained by analyzing these analogies. As noted by H. Goldstein in his book *Classical Mechanics* (Addison Wesley, Cambridge, MA, 1956), classical mechanics is only a geometrical optics approximation to a wave theory!

In this book we begin with Fermat's principle and obtain the Lagrangian and Hamiltonian pictures of ray propagation through various media. Given the current interest and activity in optical fibers and optical communication, analysis of light propagation in inhomogeneous media is dealt with in great detail. The past decade has witnessed great advances in adaptive optics and compensation for optical aberrations. The formalism described herein can be used to calculate aberrations of optical systems. Toward the end of the book, we present application of the formalism to current research problems. Of particular interest is the use of dynamic programming techniques which can be used to handle variational/extremum problems. This method has only recently been applied to optical problems.

All formulations in this book are developed from first principles, so that it can be used for self-study. Each chapter is relatively self-contained. The book is at a level suitable for the senior undergraduate/first year graduate student in Physics/Optical Sciences and Engineering/Electrical Engineering. It can also be used as an adjunct to a course in Geometric Optics. We feel that the book will help students studying Optics and/or Mechanics achieve a better understanding by working out analogies between two of the oldest disciplines in Physics.

We would like to thank the numerous colleagues and students over the years who have helped in our understanding of this subject matter. As usual all errors are ours alone. We would also like to give a special thanks to Ms. Bhawna Bharadwaj (New Delhi) and Dr. Srinivasa Varadharajan (now at the University of California at Santa Barbara) for helping us prepare a camera ready manuscript. Thanks also go to Ms. Aparna Raghuram (St. Louis) for considerable help with manuscript preparation.

The dictum "A book never gets completed, it escapes the authors" holds in this situation too. To quote Leonhard Euler (*Methodus Inveniendi Lineas Curvas*, 1744), " All the greatest mathematicians have long since recognized that the method presented in this book is not only extremely useful in analysis, but that it also contributes greatly to the solution of physical problems". Euler (1707-1783) was referring to the calculus of variations, which is the natural language of all variational principles. We hope the same applies to the present volume. May the reader enjoy and benefit!

Vasudevan Lakshminarayanan
Ajoy K. Ghatak
K. Thyagarajan

St. Louis and New Delhi

1. INTRODUCTION

Variational principles are of great universality and play a major role in the description of natural phenomena. The minimum hypothesis to explain nature has been present in ideas presented by early Greeks whose philosophy seems to be based on the principle that "Nature does nothing in vain". Such 'economy' postulates in the hands of later scientists acted as the foundation for the development of minimum principles such as Fermat's principle. If one studies the chronological development in the evolution of minimum principles one can get a profound insight into the continuos transformation of a metaphysical canon into an exact natural law.

Starting with the work of Fermat, Euler and Lagrange physicists have succeeded in formulating laws of elasticity, electrodynamics, hydrodynamics, gravitation as well as optics in terms of variational principles. Even the laws of quantum mechanics and quantum field theory can be expressed in terms of variational methods and the action principle of mechanics. It is not an exaggeration to state that variational principles play a fundamental role.

A historical review

Ray optics can be derived from Fermat's principle of least time. Classical dynamics obey not only Newton's law of motion but also the Principle of Least Action. The development of these ideas goes back to the time of the ancient Greeks when Hero (~125 B.C.E) proposed the following problem : how do you determine the angles formed by the incident and reflected light at a mirror given that light " strives to move over the shortest possible distance, since it has not the time for slower motion". This assumption is a statement of what is called the final cause. The final cause (that light minimizes its travel time between two points) of an effect (the light rays path) is that for which the effect is achieved. Since the time of Aristotle (~350 B.C.E) natural laws have been ascribed to a final cause. Even though this metaphysical "final cause" might seem strange to modern readers variational principles such a Fermat's principle are contemporary descendents' of final cause [see reference 1 for an interesting discussion] and the calculus of variations is the natural language of variational principles. Practically every law in physics can be derived from a variational principle.

Variational principles are simple statements of natural law, which have wide application and can be used to solve complex problems easily. These forms serve as a template out of which new laws can be created. For historical and critical surveys the reader is referred to references 2 and 3. Such laws and physical principles organize our observations and experimental results. According to the Physicist and Philosopher Ernst Mach, " the principle does not so much promote our insight into true processes as it secures us a practical mastery of them. The value of the principle is of an economical character" [4]. In this book we are concerned with ray optics and particle dynamics both of which display a structural similarity.

In 1675 the astronomer Olaf Roemer (1644-1710) based on his observations of I_o, a moon of Jupiter established the fact that light propagates with a finite speed. Willebrod Snell (1591-1626) found experimentally, in about 1621, that the sine of the angle of incidence and refraction form a ratio with each other which depends only on the two media involved. This is known today as Snell's law of refraction. Pierre de Fermat (1601-1665) in 1661, showed that light rays travel along the quickest paths in the two media. Fermat was following Rene Descartes (1596-1650) in a theoretical study of the motion of light by conceptually replacing a light ray whose speed was throught to be infinite (an idea expounded by Johann Kepler and formed the existing view of the speed of light) by a moving projectile whose finite speed would change upon going from one medium to another. Hence, the idea of a finite speed for light was introduced even before the result due to Roemer. By assuming that the speed of light changes in going from one medium to another and that light takes the quickest path, Fermat was able to derive Snell's law. He proposed the general principle that "

Nature always acts by the quickest time". Slightly correcting the original formulation as given by Pierre de Fermat in his Oevreus, (1891)" Je reconnius premie'rement ...la ve'rite' de ce principe Que la nature agit toujours parles voies les plus courtes", Fermat's principle states that among all possible paths connecting two given points P_1 and P_2 , a light ray would follow the path for which the optical path length between the two points is an extremum, i.e.,

$$\delta L = \delta \int_{P_1}^{P_2} n(x, y, z)\, ds = 0 \tag{1}$$

The δ variation can be understood as a variation of the path of the integral with the end points P_1 and P_2 being fixed. To be an extremum means that the optical path length may be a minimum (the case most often encountered), a maximum or stationary. In the above equation, we have taken n(x,y,z) to be the refractive index function for the medium accounting for it's optical properties and ds is an infinitesimal arc length measured along the path. It should be pointed out that in some situations such as reflection by a concave spherical mirror such as the inside of a reflecting globe, the ray of light actually follows the path of greatest time among all such admissible paths having a single reflection at the mirror. Fermat himself was aware of difficulties involving reflection of light rays by curved mirrors. This example of the concave spherical mirror is given by Luneburg [5, see also reference 6]. A fascinating account of the development of optics can be found in the books by Whittaker [7] or Ronchi [8].

Fermat's hypothesis that light travels more slowly in an optically dense medium, and indirectly the principle of least time was verified in 1850 when Leon Foucault measured the speed of light in water and found it to be less that of air by exactly the same amount predicted by Fermat's principle.

A similar situation is found in mechanics. The principle of least action determines the trajectory of a particle by extremizing the trajectory's action. The French mathematician Pierre Louis Moreau de Maupertuis who in 1740 enunciated the principle which states that the "action" is the quantity that is minimized by the actual trajectory of a mechanical system. The action is simply the product of the mass (m) times the velocity (v) and of all possible paths between P_1 and P_2, The actual path traversed by the particle's one for which

$$\delta S = \delta \int_{P_1}^{P_2} mvds = 0 \tag{2}$$

Here, we of course, impose the condition that the particle's energy is a constant of motion. It is interesting to note that Maupertuis claimed that his principle applied not only to particles but also to light rays. This attempt to incorporate ray optics into mechanics was not new at that time. Issac Newton (1642-1727) as well as Descartes had tried to do so. For example, Descartes analyzed the dynamics of cannon balls shot into the water and showed that they "reflect" and "refract" similar to light rays. He writes, "this has sometimes been demonstrated with unfortunate consequences when someone firing guns for fun into the bed of the river has wounded those who were on the other side of the bank" (from the second discourse of Descartes's Dioptrique, quoted in page 117 of reference 7).

The principle of least action was first published as an exact dynamical theorem by Leonard Euler (1703-1783) who in 1744 proved it for a single particle moving in a plane. Mauerpertuis who believed the principle reflected the simplicity and economy of God's work ("It is what I call the principle of the least quantity of action, a principle that is so wise and so worthy of the Supreme Being, and to which nature appears to be so constantly subject that she observes it not only in all her changes but still tends to observe it in her permanence"), never gave the principle a precise mathematical interpretation. It was Carl Gustav Jacob Jacobi (1804-1851) who showed that the action of a particle trajectory is

$$A = \int \sqrt{2m\{E - V(x, y, z)\}}ds \tag{3}$$

where E represents the total energy and V is the potential energy and

$$E = T + V(x, y, z) = \text{cons tan t}. \tag{4}$$

Here T is the kinetic energy given by ½ mv^2. This principle is known a Jacobi's principle.

It was Joseph Louis Lagrange (1736-1813) who at the age of 19 conceived of and at the age of 23 formulated an ingenious analytic method which allowed him to generalize D'Alembert's principle in mechanics. Working directly from the principle of least action, Lagrange developed a general variational method in dynamics. He was able to develop the method of Lagrange multipliers, a systematic approach to generalized coordinates, and analysis of many particle systems. His book, "Mechanique Analytique" contained not a single diagram, or a construction or geometric or mechanical reasoning. The book was just full of algebraic operations. E.T. Bell called it " the finest example in all science of the art of getting something out of nothing"[9]. William Rowan Hamilton (vide infra) has said " Lagrange has perhaps done more than any other analyst, to give extent and harmony to such deductive researchers by showing that the most varied consequences respecting the motion of systems of bodies may be derived from one radical formula: the beauty of the method so suiting the dignity of the results as to make his great work a kind of scientific poem"[10].

Of particular importance in our context is the idea of the Lagrangian, defined as

$$L = T - V \tag{5}$$

From a knowledge of the Lagrangian of a system, it is possible to derive it's equations if motion. The chief disadvantage of the Lagrangian formalism is that the so-called Euler-Lagrange equations are second order total differential equations in generalized co-ordinates.

The next great advance comes from the work of Wiliam Rowan Hamilton(1805-1865). Hamilton by the age of 21 remodeled geometrical optics and mechanics. In his paper " Theory of Systems and Rays" published in 1828 in Transactions of the Royal Irish Academy as well as in subsequent papers on the same journal in 1830 and 1837, he addressed the problem of studying the geometry of light rays as they pass through some optical system. In 1834 he published a short note in the British Association Report entitled " On the application to dynamics of a general mathematical method previously applied to optics". In the same year, but in the more prestigious journal,

Philosophic Transactions of the Royal Society, appeared his masterpiece "On a general method in dynamics by which the study of the motion of all free systems of attracting or repelling points is reduced to the search and differentiation of one central relation or characteristic function". Here Hamilton defined the so-called Hamilton's first principal function. According to this, if a particle occupies at time t, coordinates $(x_1\ y_1)$ and at t_1 coordinates $(x_2\ y_2)$ the Hamilton's first principle function is defined to be:

$$S = \int_{t_1}^{t_2} (T - V)dt \tag{6}$$

Lanczos[3] states there is no standard name for the quantity S. Lemons[1] notes that Richard Feynmann in his Lectures on Physics Vol. II, pages 19-8 (Reading, Mass: Addison Wesley) incorrectly calls it "Action"!

Comparing equations(6) and (5) it can be seen that the integrand is nothing but the Lagrangian. Hamilton's principle of least action can be succinctly stated as

$$\delta \int_{t_1}^{t_2} L dt = 0 \tag{7}$$

In other words, a dynamical system moves from one configuration to another is such a way that the variation of the integral in equation (7) between the actual path taken and any neighboring virtual path coterminus in both space and time with the actual path is zero, or in other words is stationary. From Hamilton's principle it is possible to derive the Euler Lagrange equations (and is common). The reader is referred to the excellent text on classical mechanics by Goldstein[11] or the book on variational principles by Lemons[1].

Analogies between optics and Mechanics

We note that geometric and wave optics are intimately connected. Wave optics encompasses geometrical optics whose fundamental equation, the eikonal equation (from Greek $\varepsilon i k \bar{\omega} \upsilon$ = image) can be obtained from the wave equation in the limit of very short wavelengths (see Appendix A for a derivation). Geometrical optics is grounded in the concept of a ray of light whose trajectory is determined by the ray-equations supplemented with suitable initial values. Hamilton in his papers studied the geometry of rays of light as they pass through some optical system. He converted the problem into that of relating the local coordinates of rays entering and exiting from the system, defined with respect to the optic axis and some suitably chosen planes traverse to the axis. If the ray coordinates are suitably chosen, the input-output relations are what are called symplectic transformations (see Appendix C and Chapter 8). Symplectic transformations are traditionally called canonical transformations in physics and are generated by a function of the ray variables, the so called characteristic function whose functional form is determined by the physical properties of the optical system. From classical mechanics it is well known that the phase space (space formed by the generalized coordinates and momenta) is a symplectic manifold and the time evolution of a mechanical system can be described in terms of canonical transformations generated by the Hamiltonian function of the system. The analogy between geometric optics and mechanics can be obtained if we replace the optic axis by the time axis, light rays by the trajectories of the particles and the ray-coordinates by the phase space variables appropriate to the mechanical system under study. In fact, since Hamilton's optical work was unknown in Europe, Bruns in 1895 suggested that the well-established Jacobi theory of mechanics be applied to geometric optics and identifies a sort of characteristic function called the eikonal [12]. In fact, the characteristic functions of Hamilton are themselves often referred to as Eikonals. It is easy to see that if one compares the Fermat's principle and Jacobi's principle, n (x,y,z) is proportional to E-V(x,y,z). The resulting Euler Lagrange equations are mathematically identical. Even though the proportionality between the two is physically difficult to conceive of, the fact is that for every ray path there is a particle trajectory and vice versa.

Without going into detail, it can be shown that a quantity W called Hamilton's characteristic function (which generates a canonical transformation) plays the same role as the eikonal and $[2m(E-V)]^{1\backslash 2}$ serves as the index of refraction. As noted by Goldstein[11], classical mechanics corresponds to the geometric optics limit of a wave motion. Corresponding to the wave amplitude ϕ in the wave equation

$$\nabla^2\phi - \frac{1}{u^2}\frac{d^2\phi}{\partial t^2} = 0 \qquad (8)$$

where u is the wave velocity in medium of index n, there will be some quantity ψ in the wave theory of mechanics which will satisfy an equation of the same form. In fact, the wave equation for which W represents the eikonal is nothing but the Schroedinger equation of quantum mechanics! If Hamilton had gone a bit further he would have discovered the Schroedinger equation, but he did not since such a leap required experimental evidence which did not exist in his time. A detailed analysis of ray optics-particle mechanic analogy is given in the paper by Evans and Rosenquist [13].

It has been shown that in full analogy with quantum mechanics, the wave optics can be regarded as the quantum theory of light rays resulting from Hamiltonian optics according to the quantization rules of the quantum-classical correspondence principle restated in optics language [14]. In this regards the wave equation is equivalent to the Klein Gordon equation of relativistic quantum mechanics and the parabolic wave equation, basic to paraxial optics is equivalent to the Schroedinger equation of non-relativistic quantum mechanics. The theory of particle optics originated in 1926 when Hans Busch showed that the action of a short rotationally symmetric magnetic field on electron rays was similar to that of a glass lens on light rays bringing electrons from an object point together at an image point [15]. In fact the optics of charged particle beams is a striking example of the formal similarity between classical mechanics and geometrical optics. It is our objective in this book to exploit these similarities and study a number of interesting problems occurring in optics.

Lagrangian and Hamiltonian Formulations in Optics – A general View

Knowing Fermat's principle and Hamilton's principle it is easy to write down the general formulations of an optical problem in terms of mechanics.

Given two points P_1 and P_2 and a curve C connecting them, the geometric path length l between P_1 and P_2 is defined as the length of the portion of C lying between these 2 points.

$$C = \int_{P_1}^{P_2} c\,ds \tag{9}$$

The corresponding optical path length is given by

$$l = \int_{P_1}^{P_2} cn(x, y, z)ds \tag{10}$$

As before, n(x,y,z) is the refractive index function and for a homogeneous isotropic medium, n(x,y,z) = n_0, the optical path length is just the geometric path length multiplied by the refractive index.

From Fermat's principle we know that

$$\delta l = \delta \int_{P_1}^{P_2} L(x_1\ y_1, x'\ y'\ z')dz = 0 \tag{11}$$

The infinitesimal arc length ds is given by

$$ds = dz\sqrt{1+x'^2+y'^2} \tag{12}$$

with $x' = \frac{dx}{dz}; y' = \frac{dy}{dz}$

and we can rewrite Fermat's principle in the same form as Hamilton's principle (equation (7))

$$\delta \int_{P_1}^{P_2} L(s, y, x', y', z)dz = 0 \tag{13}$$

and identify the integrand as the optical Lagrangian

$$L(x, y, x', y', z) \equiv n(x, y, z)\sqrt{1+x'^2+y'^2} \tag{14}$$

Here z plays the role as time in Lagrangian mechanics and is an independent variable. Usually the z direction is chosen to be the optic axis of the system. The Lagrangian equations are obtained as solutions to the variational principle. Equation (13) can be written as :

$$\frac{\partial}{\partial z}\left(\frac{\partial L}{\partial x'}\right)-\frac{\partial L}{\partial x}=0$$
$$\frac{\partial}{\partial z}\left(\frac{\partial L}{\partial y'}\right)-\frac{\partial L}{\partial y}=0 \tag{15}$$

The above equations can be written in terms of x,y and n as

$$\frac{d}{ds}\left(n\frac{dx}{ds}\right)=\frac{\partial n}{\partial x}$$
$$\frac{d}{ds}\left(n\frac{dy}{ds}\right)=\frac{\partial n}{\partial y} \tag{16}$$

A similar equation holds for the z coordinate. We can write these equations in vectors form as

$$\frac{d}{ds}\left(n\frac{d\vec{r}}{ds}\right)=\nabla n \tag{17}$$

where $\vec{r}$ is the position vector along any point on the ray path and ∇ is the gradient operator. Equation (17) is known as the ray-equation. In analogy to Hamiltonian mechanics we introduce the generalized momenta, conjugate to x and y as

$$p_x=\frac{\partial L}{\partial x'};p_y=\frac{\partial L}{\partial y'} \tag{18}$$

or explicitly

$$p_x=n\frac{x'}{\sqrt{1+x'^2+y'^2}}=n\frac{dx}{ds}$$
$$py=n\frac{y'^1}{\sqrt{1+x'^2+y'^2}}=n\frac{dy}{ds} \tag{19}$$

These equations are nothing but the local value of refractive index times the direction cosines ($\frac{dx}{ds},\frac{dy}{ds}$) of the light ray with respect to x and y axis respectively. These are termed the optical direction cosines. Considering these equations it is easy to see that refractive index n turns the geometric variables ($\frac{dx}{ds},\frac{dy}{ds}$) into the canonically conjugate variables p_x, p_y. In mechanics we replace velocity with momentum, the mass being the multiplicative factor. The Hamiltonian for the ray is given by the well know procedure –

$$H(x, y, p_x, p_y, z) = p_x x' + p_y y' - L \tag{20}$$

Substituting for variables , we get

$$H(x, y, p_x, p_y, z) = -\sqrt{n^2 - p_x{}^2 - p_y{}^2} \tag{21}$$

and the equations of motion are given by

$$\frac{\partial q_x}{\partial z} = \frac{\partial H}{\partial p_x} \qquad \frac{\partial q_y}{\partial z} = \frac{\partial H}{\partial p_y}$$

$$\frac{\partial p_x}{\partial z} = -\frac{\partial H}{\partial q_x} \qquad \frac{\partial p_y}{\partial z} = -\frac{\partial H}{\partial q_y} \tag{22}$$

where we have made the simple substitution $x \to q_x, y \to q_y$. Identical expression hold in mechanics.

Given the initial values of the coordinates (q_x, q_y) and optical direction cosines (p_x, p_y) at same initial value of the equation of motion, it is possible in general to solve Hamilton's equation for ray position and momenta at any z. These equations provide the formal answer to the basic problem in geometrical optics of determining the final coordinates of the ray after passing through an optical system, given the initial ray coordinates and the optical properties of the system.

Lagrangian versus Hamiltonian Optics

Buchadahl in his classic book on Hamiltonian optics [16],has discussed the issue of which method is preferred in optical computation. The problem can be classified into two different techniques:

(1) we can specify two points, one in object space and the other in image space and determine the directions of the ray passing through the points or
(2) Selects a point on a ray in object space and its direction and determine in image space a point on the ray and its direction.

The analogous situation in dynamics is :

(1) If a particle P were at a point $(x_1y_1z_1)$ at time t_1 and at $(x_2\ y_2\ z_2)$ at time t_2 what were the components of momentum at those two points?

Alternatively, the problem can be recast as

(2) if the particle were at $(x_1y_1z_1)$ at time t_1 with momentum p_1, where will the particle be at time t_2 and what will its momentum p_2 be ?

For either optical case or dynamical case question (1) leads to the determination of the so-called point characteristic which is the analog of the Hamilton's principal function. For the second question, where the initial configuration and state of motion are given, then the problem can be solved using Lagrange's equations. In the optical case, we have to solve, for a homogenous isotropic medium, the ray equation. From a purely computational point of view Lagrange's method is probably simpler when considered with the dynamics analogy. Also, it is easier to work with object space coordinates. To quote Buchdahl " The great simplicity of the Lagrangian method of computation leads me to believe that it rather than the Hamiltonian method is best adapted to the problem of practical calculations: though this is admittedly a mere expression of opinion".

Given the power of modern digital computers, the choice of which method to use is simply one of convenience and preference. Sufficient number of examples of these techniques are given in this text. It should be pointed out that phase space formulation of quantum and classical mechanics is becoming increasing popular [17]. These methods have been used to study transport of charged particle beams and has been used to deal with a number of problems in classical and quantum optics (e.g., reference 18). Phase space methods seem to be an appropriate technique to transport problems and they have been applied to light optics to provide an overall context where both wave and geometrical optics can be framed. Amalia Torre [19] has provided a unified point of view to radiation and charged beam transport, and is currently working on a book on phase space methods in optics (private communication).

As noted previously the Hamiltonian nature of the ray equations of motion assumes that the ray variables at a given z (q_i, p_i) are related to (q_f, p_f) through a symplectic transformation. The Hamiltonian technique in optics concerns solutions of Hamilton's equations of motion for the phase space variables (q, p)

$$\frac{d}{ds}\begin{pmatrix} q \\ p \end{pmatrix} = -\left\{ H, \begin{pmatrix} q \\ p \end{pmatrix} \right\} \tag{23}$$

where {,} represents the classical Poisson bracket. When dealing with intense beams, it is possible to deal with concepts like phase space distribution function $\rho(q,p;s)$, beam emittance, brightness. etc. These methods take into account a collective or statistical description instead of single rays, in which case one deals with the Liouville's equation

$$\frac{d}{ds}\rho = \{H, \rho\} \tag{24}$$

It can be stated that Hamiltonian optics deals with equation of the form

$$\frac{d}{ds}v = \widehat{H}v \tag{25}$$

where v denotes the phase space vector (q,p) or the phase space distribution function $\rho(q,p,s)$ and $\widehat{H}$ is the operator $\pm\{H,\cdot\}$ generated by the Poisson bracket. It is known that in general the symplectic maps can be generated using the Hamiltonian (see chapter 8). In this context, Hamiltonian methods seem to be the better choice. These techniques are of current research interest and further discussion is beyond the scope of this book.

References:

1. D.S. Lemons, *Perfect form*, Princeton University Press, Princeton, N J (1997)
2. W. Yourgrau and S. Madelstam, *Variational principles in Dynamics and Quantum theory*, Dover, NY, (1968)
3. C. Lanczos, *The variational principles in Mechanics*, Dover, NY, (1970)
4. E. Mach, *The science of mechanics*, Open court: La Salle, Illinois, (1960). Page 30
5. R.K. Luneburg, *Mathematical theory of optics*, Univ. of California press, Berkeley, CA. (1964) page 87.
6. C. Caratheodory, *Geometrische optik*, Springer Verlag, Berlin, 1937. page 10.
7. V. Ronchi, *The nature of light*, Harvard University Press, Cambridge, MA (1970)
8. E.T. Whittaker, *A history of the theories of Aether and Electricity*, Thomas Nelson and Sons, London Vol. 1 (1951); Vol .2 (1953)
9. E.T. Bell, *Men of Mathematics*, Simon and Schuster, NY, 1937

10. W.R. Hamilton, ***The mathematical papers of Sir William Rowan Hamilton***, ed. A.W. Conway and J.L. Synge, Vol. 1: Geometrical optics, Cambridge Univ. Press, London (1931); Vol .2: Dynamics, Cambridge Univ. Press, London, 1940

11. H.Goldstein, ***Classical Mechanics***, Addison Wesley, Cambridge, Mass., (1956)

12. H. Bruns, Abh. Kgl. Sachs. Ges.Wiss., Math-phys. KL, 21,323 (1985); see also J.L. Synge, J. Opt. Soc. Am., 27:138, (1937)

13. J.Evans and M.Rosenquist, ***'F=ma optics'***, Am. J. Phys. 54:876-883 (1986)

14. D. Gloge, D. Marcuse, ***Formal quantum theory of light rays***, J. Opt. Soc. Am., 59:1629, (1969)

15. H. Busch, Zeitschr. Physik, 81,:974 (1926)

16. H.A. Buchadahl, ***An introduction to Hamiltonian optics***, Dover, NY, (1993)

17. Y.S. Kim and M.S. Noz, ***Phase space picture of Quantum mechanics***, world scientific, Singapore (1991)

18. C. Lejeure and J. Aubert, in ***Applied charged particle optics***, ed. A.Septier, Adv. in Electron. and electr. Physics suppl. 13A, Academic press, NY 1980.

19. A.Torre, ***A unified point of view to radiation and changed beam transport***, Il Nuovo Cimento, 112: 409 (1999)

2. FERMAT'S PRINCIPLE

Introduction

As mentioned in chapter 1, optics and mechanics have a number of analogies which have been used time and again to understand and appreciate concepts in either of the fields. One of the most basic analogy is between the Fermat's principle in optics and the Hamilton's principle of least action in classical mechanics [1]. As in classical mechanics, we can use either the Lagrangian or the Hamiltonian formulations to further study properties of light ray propagation. We will first develop the Lagrangian formulation which will be followed by the Hamiltonian formulation .

Before we introduce Fermat's principle, it is necessary to introduce the concept of optical pathlength. Given any two points *P* and *Q* and a curve *C* connecting them, one can define the geometrical pathlength between the two points as the length of the curve lying between the two points, $\int_P^Q ds$, where the integral is performed from *P* and *Q* along the curve *C* and *ds* represents an infinitesimal arclength (Fig. 2.1). The optical pathlength is defined as

$$\text{optical path length} = \int_P^Q n(x,y,z)ds \tag{1}$$

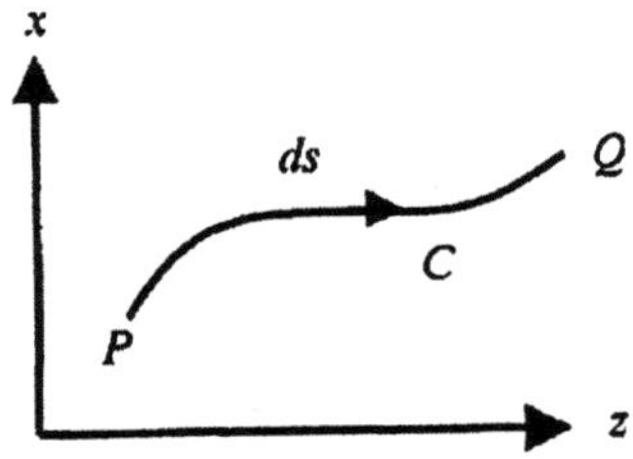

Figure 2.1 Curve C connects points P and Q.

where $n(x, y, z)$is the space dependent refractive index function and the integral is again performed along the curve C. In the simple case of a homogeneous medium, the optical pathlength is just the geometrical pathlength multiplied by the refractive index of the medium. In the general case, the optical pathlength divided by c (the velocity of light in free space) represents the time that would be required for light to travel from P to Q along the given curve.

We may now state Fermat's principle according to which, out of the many paths that can connect two given points P and Q, the light ray would follow that path for which the optical pathlength between the two points is an extremum, i.e.,

$$\delta \int_{P}^{Q} n(x,y,z)ds = 0 \tag{2}$$

where the delta variation of the integral means that it is a variation of the path of the integral such that the endpoints P and Q are fixed. It should be noted that Fermat's principle requires the optical pathlength to be an extremum, which may be a minimum (this is the case one most often encounters), a maximum, or stationary with respect to variations in the path keeping the end points P and Q fixed. It is at once clear that in a homogeneous medium, the rays are straight lines, since the shortest optical pathlength between two points is along a straight line.

It is of historical interest to note that Hero of Alexandria (125 B.C.) was the first person to solve the so called "Hero's problem" and derived the law of reflection by using a minimum principle. According to Hero, light "strives, to move over the shortest possible distance since it has not the time for slower motion" [2,3].

We give below some illustrative applications of Fermat's principle. (Fermat's principle for an anisotropic medium is discussed in appendix B.)

Example 1:

We first obtain the law of reflection using Fermat's principle. If a ray begins to travel from a point $A(x_1, y_1)$ above a flat reflecting surface and travels in a straight line to the surface at a point $C(x, 0)$, reflects and returns to a different point at $B(x_2, y_2)$, then we can write, (Fig. 2.2) for the time taken to travel from point A to point B as (assuming the velocity of light to be c in the medium)

$$T(x) = \frac{1}{c}\left[\sqrt{(x-x_1)^2 + y_1^2} + \sqrt{(x_2-x)^2 + y_2^2}\right] \tag{3}$$

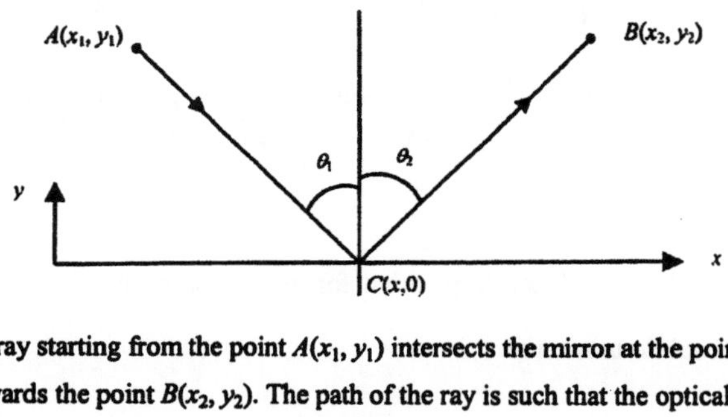

Figure 2.2 A ray starting from the point $A(x_1, y_1)$ intersects the mirror at the point $C(x, 0)$ and proceeds towards the point $B(x_2, y_2)$. The path of the ray is such that the optical path length $AC+CB$ is a minimum.

The law of reflection can be derived using Fermat's principle by solving the equation

$$\frac{\partial T(x)}{\partial x} = 0 \tag{4}$$

which implies that

$$\frac{x-x_1}{\sqrt{(x-x_1)^2+y_1^2}}=\frac{x_2-x}{\sqrt{(x_2-x)^2+y_2^2}}$$

from which it is easy to see that

$$\sin\theta_1=\sin\theta_2$$

or

$$\theta_1=\theta_2 \tag{5}$$

where θ_1 and θ_2 are the angles of incidence and reflection (Fig. 2.2). This is simply the law of reflection. It is an easy exercise to show that this is a true minimum. For,

$$c\frac{\partial^2 T}{\partial x^2}=cT''(x)=\frac{1}{\sqrt{(x-x_1)^2+y_1^2}}[1-\sin^2\theta_1]+\frac{1}{\sqrt{(x_2-x)^2+y_2^2}}[1-\sin^2\theta_2]$$

$$=(1-\sin^2\theta_2)\left[\frac{1}{\sqrt{(x-x_1)^2+y_1^2}}+\frac{1}{\sqrt{(x_2-x)^2+y_2^2}}\right]$$

Thus, since $\sin^2\theta_2 \leq 1$, $T''(x)$ is positive for all values of x. Therefore, $T(x)$ at $T'(x)=0$ is an absolute minimum.

Example 2

To obtain the laws of refraction, let *MN* be the surface separating two media of refractive indices n_1 and n_2 (Fig 2.3). Let the ray start from A, intersect the surface at *C* and proceed to *B* along *CB*. *AC* and *CB* must be straight lines since they are in homogeneous media. Let *P* and *Q* represent the feet of the perpendiculars from *A* and *B* to the surface and let $PQ = L$. Let x be the distance *PC*. We have to find the point *C* such that the optical pathlength *ACB* is a minimum. The optical pathlength of *ACB* is

$$\Lambda = n_1\sqrt{d_1{}^2 + x^2} + n_2\sqrt{d_2{}^2 + (L-x)^2} \qquad (6)$$

where $d_1 = AP$ and $d_2 = BQ$. For Λ to be an extremum with respect to x, we must have

$$\frac{d\Lambda}{dx} = n_1\frac{x}{\sqrt{d_1^2 + x^2}} - n_2\frac{(L-x)}{\sqrt{d_2^2 + (L-x)^2}} = 0$$

If θ_1 and θ_2 are the angles defined as in Fig.2.3, then

$$\sin\theta_1 = \frac{x}{\sqrt{d_1^2 + x^2}}$$
$$\sin\theta_2 = \frac{(L-x)}{\sqrt{d_2^2 + (L-x)^2}} \qquad (7)$$

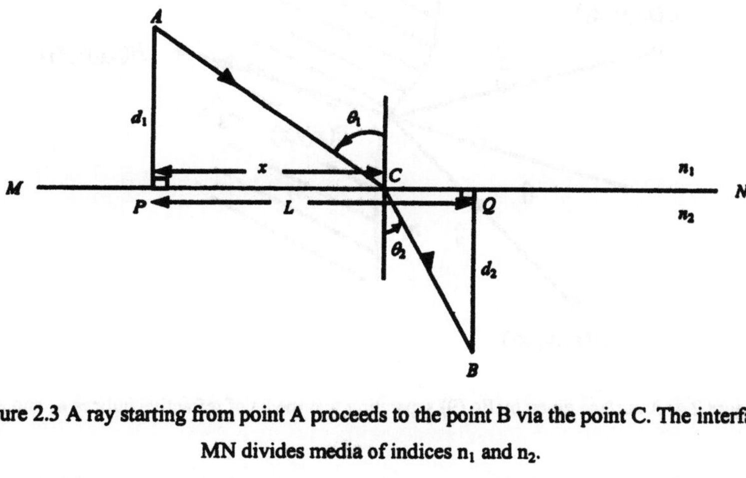

Figure 2.3 A ray starting from point A proceeds to the point B via the point C. The interface MN divides media of indices n_1 and n_2.

Thus, for the optical path to be an extremum, we must have

$$n_1 \sin\theta_1 = n_2 \sin\theta_2 \qquad (8)$$

which is Snell's law. This path can again be shown to correspond to an absolute minimum.

Example 3

We next generalize the above results when the surface of reflection or refraction is not plane but is given by an equation of the form

$$f(x, y, z) = 0 \tag{9}$$

The analysis will follow the one given in Ref. 4.

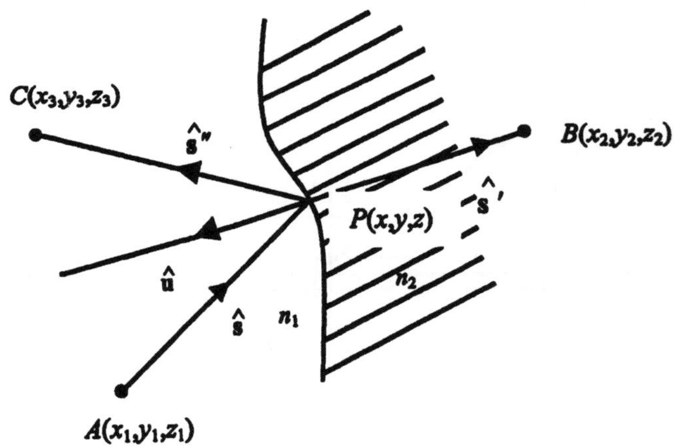

Figure 2.4 A surface given by Eq.(9) separating two media of refractive indices n_1 and n_2.

Let the surface given by Eq.(9) separate the two media of refraction (see Fig.2.4). Let a ray starting from $A(x_1, y_1, z_1)$ reach the point $B(x_2, y_2, z_2)$ after getting refracted from the surface at a point $P(x, y, z)$. Let $\hat{\mathbf{u}}$ be the unit normal at P to the surface. Let $\hat{\mathbf{s}}$

and $\hat{s}'$ be the unit vectors along *AP* and *PB* respectively. We have to find a relation between $\hat{s}$, $\hat{s}'$ and $\hat{u}$. The optical pathlength Λ (between *A* and *B*) along *APB* is

$$\Lambda(x,y,z) = n_1 d_1 + n_2 d_2 \tag{10}$$

where

$$d_1 = \sqrt{(x-x_1)^2 + (y-y_1)^2 + (z-z_1)^2} \tag{11}$$

$$d_2 = \sqrt{(x-x_2)^2 + (y-y_2)^2 + (z-z_2)^2} \tag{12}$$

If the point $P(x, y, z)$ changes to $P_1(x + \delta x, y + \delta y, z + \delta z)$ then the change in Λ is given by

$$\delta\Lambda = (n_1\alpha_1 - n_2\alpha_2)\delta x + (n_1\beta_1 - n_2\beta_2)\delta y + (n_1\gamma_1 - n_2\gamma_2)\delta z \tag{13}$$

where $\alpha_1[= (x - x_1)/d_1]$, $\beta_1[= (y - y_1)/d_1]$, $\gamma_1[= (z - z_1)/d_1]$ and α_2, β_2, γ_2 (with similar definitions) represent the direction cosines of *AP* and *PB*, respectively, i.e., they are the *x*, *y*, and *z* components of $\hat{s}$ and $\hat{s}'$. Further, in Eq.(13) δx, δy and δz cannot be varied arbitrarily because the changes δx, δy, and δz have to be such that the point *P* still lies on the surface given by Eq.(9); thus we must have

$$\frac{\partial f}{\partial x}\delta x + \frac{\partial f}{\partial y}\delta y + \frac{\partial f}{\partial z}\delta z = 0 \tag{14}$$

If we substitute for δz from Eq.(14) into Eq.(13), we obtain an equation in which δx and δy can be varied arbitrarily. As such, the coefficients of δx and δy must be set equal to zero, which would lead to

$$\frac{n_1\alpha_1 - n_2\alpha_2}{\partial f/\partial x} = \frac{n_1\beta_1 - n_2\beta_2}{\partial f/\partial y} = \frac{n_1\gamma_1 - n_2\gamma_2}{\partial f/\partial z} = K \tag{15}$$

where K is a constant. This yields

$$n_2\alpha_2 - n_2\alpha_2 = K\frac{\partial f}{\partial x} \tag{16}$$

etc. Observing that the vector with components $\partial f/\partial x$, $\partial f/\partial y$ and $\partial f/\partial z$ represents[1] the direction of the normal to the surface ($\hat{\mathbf{u}}$), we can write Eq.(15) as

$$n_1\,\hat{\mathbf{s}} - n_2\,\hat{\mathbf{s}}' = K_1\,\hat{\mathbf{u}} \tag{17}$$

where K_1 is K times the magnitude of the vector with components $\partial f/\partial x$, $\partial f/\partial y$ and $\partial f/\partial z$. Equation (17) implies that $\hat{\mathbf{s}}$, $\hat{\mathbf{s}}'$ and $\hat{\mathbf{u}}$ are coplanar, i.e., the incident ray, the refracted ray, and the normal to the surface lie in the same plane. Taking a cross product of Eq. (17) with $\hat{\mathbf{u}}$ we get

$$n_1 \sin i_1 = n_2 \sin i_2 \tag{18}$$

where i_1 is the angle between $\hat{\mathbf{u}}$ and $\hat{\mathbf{s}}$ and i_2 is the angle between $\hat{\mathbf{u}}$ and $\hat{\mathbf{s}}'$. Equation (18) is Snell's law. In a similar manner, one can obtain the laws of reflection.

---

[1] The normal to a surface $f(x,y,z)$ = constant is given by ∇f, and hence the components are $\partial f/\partial x$, $\partial f/\partial y$ and $\partial f/\partial z$.

Example 4

Next, we derive the Gaussian thin lens and mirror formulas from Fermat's principle. Reference 5 has derived these using geometrical methods using only the image-forming, symmetry and ray-redirecting properties of an abstract "optical element". Here, the Gaussian formula relating object distance, u, image distance v and focal length f of a thin lens or mirror along the axis

$$\frac{1}{u}+\frac{1}{v}=\frac{1}{f} \tag{19}$$

will be derived using Fermat's principle and analytic methods without specifying a particular lens or mirror geometry in the most general formula. Referring to Fig.2.5, we show the geometry and the positive values of the coordinates (u, v, y_0, y_i) and h, the height above the optical axis the ray intersects the first surface of the optical element. (u, y_0) represents the coordinates of the object point and (v, y_i) the coordinates of the image point, and let $T(h)$ represent the time delay of the ray due to the optical element. We assume an optical element that is thin and axially symmetric. This assumption implies that the time delay $T(h)$ is only dependent on the height h above the symmetry axis (or optical axis). This also implies that all odd-derivatives of $T(h)$ vanish at $h = 0$ (due to axial symmetry).

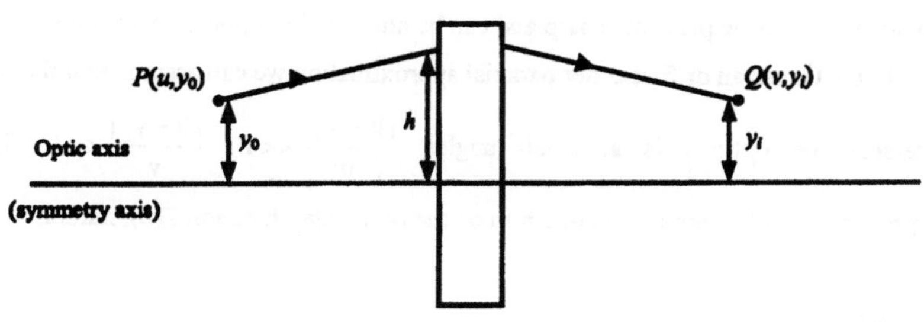

Fig. 2.5 A ray starting from the point P is deviated by an optical element and then passes through a point Q. If all ray emanating from P pass through Q, then Q will correspond to the image of P.

The optical pathlength is simply

$$\Lambda = \sqrt{u^2 + (h - y_0)^2} + \sqrt{v^2 + (h - y_i)^2} + cT(h) \tag{20}$$

where c is the velocity of light in the optical element. By Fermat's principle, the ray path is one that makes Λ stationary with respect to h. But, the caveat is that for sharp imaging, all rays which intersect the optical element from the object must come to a focus at the same image point. Taken together, the two conditions require Λ to be independent of h. In general, the function $v(u, y_0)$ and $y_i(u, y_0)$ which determine the image position in terms of the object position satisfying the constraints are impossible to find. This can be seen by the fact that if we express Λ in a power series of h

$$\Lambda = [\Lambda_0] + \frac{h}{1!}[\Lambda_0'] + \frac{h^2}{2!}[\Lambda_0''] + \cdots \tag{21}$$

where the primes indicate differentiation with respect to h, evaluated at $h = 0$. Since the powers of h are linearly independent, the coefficient of each non-zero power must vanish, i.e., $\Lambda_0^n = 0$ for $n = 1, 2, 3 \ldots$

The derivatives of Λ as given above are not identically zero, such that the two unknowns v and y_i cannot satisfy an infinite number of non-trivial relation and hence $v(u, y_0)$ and $y_i(u, y_0)$ cannot exist. What does this mean physically? Simply stated the image will never be perfectly sharp and can be attributed to optical aberrations.

In the Gaussian or first-order paraxial approximation we can assume that the rays intersect the optic axis at small angles $\frac{|h - y_0|}{u} \ll 1$, $\frac{|h - y_i|}{v} \ll 1$. This approximation also imposes a condition on the time delay function $T(h)$, namely

$$\frac{h^n}{n!} T^n(0) \ll u^n \sim v^n \tag{22}$$

If this is satisfied, we can assume that y_0, y_i and h are small so that

$$\Lambda = u\left[1+\frac{(y_0-h)^2}{2u^2}\right]+v\left[1+\frac{(y_i-h)^2}{2v^2}\right]+c\left[T(0)+T'(0)h+T''(0)\frac{h^2}{2}\right]$$
$$=\left[u+v+cT(0)\right]+h\left[-\frac{y_0}{u}-\frac{y_i}{v}+cT'(0)\right]+\frac{h^2}{2}\left[\frac{1}{u}+\frac{1}{v}+cT''(0)\right] \qquad (23)$$

By Fermat's principle, the right hand side of the above equation must be independent of h if the optical element forms sharp images. This means that the coefficients of h and h^2 are zero. That is (since $T'(0) = 0$)

$$\frac{y_i}{y_0}=\frac{v}{u} \qquad (24)$$

$$\frac{1}{u}+\frac{1}{v}=-cT''(0) \qquad (25)$$

Since v is determined by u, without a y_0 dependence, it is possible for image points (and planes) to exist. When u is at optical infinity by definition the image position will be at the focal length. In this limit, we can identify the right hand side of Eq.(25) to be equal to $-1/f$. With this identification, it is easy to see that Eq.(25) is nothing but the usual thin lens formula and Eq.(24) is the formula for lateral magnification. If we keep the higher order terms in the expression for Λ, we get the aberrations. In particular, the next term in y_0 and y_i, twists the image out of the plane (astigmatism) and the next term in h characterizes spherical aberration.

How do we specify the delay function $T(h)$? A lens is nothing but a piece of transparent material whose thickness is a smooth function of distance h from the optic axis. We can also consider other delaying materials or mechanism (e.g., a non-uniform or gradient index medium), and generalize these results for variously shaped geometries and index functions. If we describe the two lens surfaces to be described by $x_1(h)$ and $x_2(h)$, then the ray is delayed by

$$T(h)=\frac{1}{c}[x_1(0)-x_1(h)+nx_1(h)]+\frac{1}{c}[x_2(0)-x_2(h)+nx_2(h)] \qquad (26)$$

$$T'(0) = \frac{n-1}{c}[x_1''(0) + x_2''(0)] \tag{27}$$

It is easy to see that $x_1''(0)$ and $x_2''(0)$ are nothing but the negative inverses of the radii of curvature of the two surfaces at $h = 0$. Therefore,

$$\frac{1}{f} = (n-1)\left(\frac{1}{R_1} + \frac{1}{R_2}\right) \tag{28}$$

The above formula is the famous "Lensmaker's equation". For a mirror given by $x(h)$,

$$T(h) = \frac{2x(h)}{c} \tag{29}$$

$$T''(0) = \frac{2x''(0)}{c} \tag{30}$$

or

$$cT''(0) = -\frac{2}{R} \tag{31}$$

and hence

$$f = \frac{R}{2} \tag{32}$$

which is the usual formula for the focal length of a spherical mirror.

Application of Fermat's principle to atmospheric refraction and the formation of mirages.

A number of interesting phenomena occur due to variation of refractive index in the atmosphere which results in curved light paths and these can be analyzed in terms of Fermat's principle [6].

If the refractive index n is a function of y (which represents the height above ground), then the optical path length connecting two points $P(x_1, y_1)$ and $Q(x_2, y_2)$ is simply

$$cT = \int_P^Q dx\, n(y)\sqrt{1 + y'^2} \tag{33}$$

where

$$y' = \frac{dy}{dx}$$

Since the integrand $f(y, y') = n(y)\sqrt{1+y'^2}$ does not depend explicitly on the independent variable x, the Euler-Lagrange equations reduce to the so-called first integral

$$f - y'(\partial f / \partial y') = \text{a constant, } k \tag{34}$$

which here is

$$\frac{n}{\sqrt{1+y'^2}} = k \tag{35}$$

k is the value of the refractive index where $y'(x) = 0$. Physically this represents the point where the ray becomes horizontal.

Because the angle θ between the ray tangent and a vertical line is related to y' by the relationship (Fig. 2.6)

$$\sin\theta = \frac{1}{\sqrt{1+y'^2}} \tag{36}$$

we can write the equation for *k* as

$$n(y)\sin\theta = k \tag{37}$$

The above equation generalizes Snell's law to a planar atmosphere where *n* is a function of the vertical coordinate. We also get a quadrature

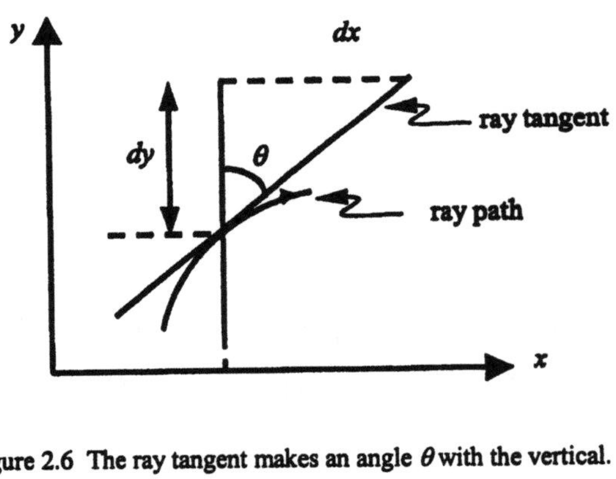

Figure 2.6 The ray tangent makes an angle θ with the vertical.

$$x - x_0 = \pm\int_{y_0}^{y} \frac{dx}{\sqrt{\left(\frac{n(y)}{k}\right)^2 - 1}} \tag{38}$$

We rewrite Eq.(35) as

$$1 + \left(\frac{dy}{dx}\right)^2 = \frac{n^2(y)}{k^2} \tag{39}$$

Differentiating the above equation with respect to *x*, we get

$$2\left(\frac{dy}{dx}\right)\frac{d^2y}{dx^2}=\frac{1}{k^2}\frac{dn^2(y)}{dy}\frac{dy}{dx}$$

or

$$\frac{d^2y}{dx^2}=\frac{n(y)}{k^2}\frac{dn(y)}{dy} \tag{40}$$

Let us consider the problem of atmospheric ray trajectory where we may assume the refraction index profile to be approximately given by the following equation. Thus, Eq.(40) gives

$$n(y)=n_0-\alpha y \tag{41}$$

$$\frac{d^2y}{dx^2}=\lambda^2 y-n_1^2 \tag{42}$$

where

$$\lambda^2=\frac{\alpha^2}{k^2}$$
$$n_1^2=\frac{n_0\alpha}{k^2} \tag{43}$$

The solution is given by

$$y(x)=c_1e^{-\lambda x}+c_2e^{\lambda x}-\frac{n_1^2}{\lambda^2} \tag{44}$$

We assume the initial values to be $y(0)=0$ and $y'(0)=0$

$$y(0) = 0 \Rightarrow c_1 + c_2 - \frac{n_1^2}{\lambda^2} = 0$$
$$y'(0) = 0 \Rightarrow -\lambda c_1 + \lambda c_2 = 0$$

Solving these two equations we get

$$c_1 = \frac{n_1^2}{2\lambda^2}$$
$$c_2 = c_1$$

Therefore, the ray trajectories are given by

$$y(x) = \frac{n_0}{2\lambda k^2}\left[e^{-\lambda x} + e^{\lambda x} - 2\right] \tag{45}$$

If $n(y)$ decreases with altitude, i.e., $\frac{dn}{dy} < 0$ the ray described by the quadrature is concave down. Light minimizes transit time by arching its path upwards between ray end points. Consequently objects are not where they seem to be but are actually a little lower than the direction in which our eyes look to see them. If we have an inversion layer where $n(y)$ increases with altitude, $\frac{dn}{dy} > 0$ and the ray is concave upward forming mirages.

Consider the example of road surface mirages. Since the hot air rises, on a hot road surface the air lower down is hotter and less dense than layers above. This is an inversion layer. Throughout the layer the pressure P is roughly constant. We know from thermodynamics that P = constant implies ρT = constant where ρ is the air density. The Gladstone-Dale law states that in a gas the deviation of n from its value in a vacuum is proportional to gas density ρ:

$$(n-1) \propto \rho$$

Let L be the distance at which rays appear to reflect from the road. Let the refractive index of hot air where the ray becomes horizontal be n_h; at eye position we let it be n_c

and the ray's angular deviation from horizontal be α. The generalized Snell's law in this case

$$n_c \sin(\pi/2-\alpha)=n_h$$
$$\cos\alpha=\frac{n_h}{n_c} \tag{46}$$

Usually $\alpha<<1$ so that $\cos\alpha \simeq 1-\alpha^2/2$ implying

$$\alpha \simeq \sqrt{2\left(1-\frac{n_h}{n_c}\right)} \tag{47}$$

Given constant air pressure

$$(n_h-1)T_h=(n_c-1)T_c$$

$$1-\frac{n_h}{n_c}=\left(1-\frac{1}{n_h}\right)\left(1-\frac{T_c}{T_h}\right)$$

$$\alpha = \text{angular deviation}$$
$$=\sqrt{2\left(1-\frac{1}{n_c}\right)\left(1-\frac{T_c}{T_h}\right)} \tag{48}$$

If on a warm day, air temperature 2 meters from the ground might be $T_c \simeq 303^0$ (30^0 C) and near road surface $T_h \simeq 323^0$ (50^0 C), from the hand book of physics, the corresponding index is $n_c = 1.00026$, and $\alpha \simeq 5.7 \times 10^{-3}$ radians. This implies that if the eye is at 2 m above the road surface, we will see a mirage at a distance $L = 2/(5.7 \times 10^{-3}) \simeq 350$ meters. The larger the temperature variation, the closer the mirage will be. This image is virtual and is the tangent to the ray at the eye which intersects the road at a distance L.

References

1. H. Goldstein, *Classical Mechanics*, Chapter 2 and Appendix A.
2. W. Yovrqrau and S. Mandelstam, *Variational Principles in Dynamics and Quantum Theory*, Dover, New York, (1979), p.5. Historical surveys can be found in this book, Chapter 1 and 14.
3. C. Lanczos, *The Variational Principle of Mechanics*, Dover, New York, 1970, Chapter X.
4. R.J. Pegis, *The modern development of Hamiltonian Optics*, in Progress in Optics (Ed. E. Wolf) Vol. I, North Holland Pub. Co., Amsterdam, (1961).
5. S. Allen, *The Gaussian formula and the elusive fourth principal ray*, Am. J. Phys. **60** (1992) 160-163. See also D simmons, *Perfer Form*, Princeton University Press, Princeton, NJ (1996) for a detailed derivation.
6. M.G.J. Minnaert, *Light and color in the outdoors*, Springer-Verlag, New York, (1993). Chapter 4 discusses the effects of curvature of light rays in the atmosphere.

3. THE OPTICAL LAGRANGIAN AND THE RAY EQUATION

Introduction

According to Hamilton's principle in classical mechanics, the trajectory of a particle between times t_1 and t_2 is such that*

$$\delta \int_{t_1}^{t_2} L(q_j, \dot{q}_j, t)\,dt = 0 \tag{1}$$

where L is called the Lagrangian, the integration is over time, q_j (j = 1,2,...) represent the generalized coordinates and dots represent differentiation with respect to time. Equation (1) is referred to as the Hamilton's principle of least action. From Eq.(1) it is possible to derive the Lagrange's equations of motion [1]:

$$\frac{d}{dt}\left(\frac{\partial L}{\partial \dot{q}_j}\right) = \frac{\partial L}{\partial q_j} \tag{2}$$

* Equation (1) is valid for conservative systems; for more details, the reader may look up Ref.1.

In this chapter we will write Fermat's principle in the form of Eq.(1) and derive the ray equation using Cartesian coordinates. We will obtain explicit solutions of the ray equation. In the next chapter we will obtain the optical Lagrangian in cylindrical coordinates and derive ray equations valid for optical fibers which are characterized by cylindrically symmetric refractive index distribution.

The ray equation in Cartesian coordinates

In the previous chapter we had studied Fermat's principle which can be written in the form

$$\delta \int_A^B n(x, y, z)ds = 0 \tag{3}$$

where $n(x, y, z)$ represents the refractive index distribution and ds represents the arc length along the path of the ray which is given by

$$(ds)^2 = (dx)^2 + (dy)^2 + (dz)^2$$

The above equation can be written in the form

$$ds = \sqrt{1 + \dot{x}^2 + \dot{y}^2}\,dz \tag{4}$$

where

$$\dot{x} = \frac{dx}{dz}, \qquad \dot{y} = \frac{dy}{dz} \tag{5}$$

Thus, Eq.(3) becomes

$$\delta \int_{A}^{B} n(x, y, z)\sqrt{1+\dot{x}^2+\dot{y}^2}\,dz = 0 \tag{6}$$

or

$$\delta \int L dz = 0 \tag{7}$$

where

$$L = n(x, y, z)\sqrt{1+\dot{x}^2+\dot{y}^2} \tag{8}$$

represents the optical Lagrangian. Since Eq.(8) is of a form identical to the Hamilton's principle in classical mechanics (Eq.(1)), we can immediately write the corresponding Lagrange's equations of motion which essentially represent path of rays in space as

$$\frac{d}{dz}\left(\frac{\partial L}{\partial \dot{x}}\right) = \frac{\partial L}{\partial x} \tag{9}$$

or

$$\frac{d}{dz}\left(\frac{n\dot{x}}{\sqrt{1+\dot{x}^2+\dot{y}^2}}\right) = \sqrt{1+\dot{x}^2+\dot{y}^2}\,\frac{\partial n}{\partial x} \tag{10}$$

Similarly,

$$\frac{d}{dz}\left(\frac{n\dot{y}}{\sqrt{1+\dot{x}^2+\dot{y}^2}}\right) = \sqrt{1+\dot{x}^2+\dot{y}^2}\,\frac{\partial n}{\partial y} \tag{11}$$

Now, Eq.(10) may be written in the form

$$\frac{1}{\sqrt{1+\dot{x}^2+\dot{y}^2}}\frac{d}{dz}\left(n\frac{1}{\sqrt{1+\dot{x}^2+\dot{y}^2}}\frac{dx}{dz}\right)=\frac{\partial n}{\partial x} \tag{12}$$

If we now use Eq.(4) we would get

$$\frac{d}{ds}\left(n\frac{dx}{ds}\right)=\frac{\partial n}{\partial x} \tag{13}$$

Similarly,

$$\frac{d}{ds}\left(n\frac{dy}{ds}\right)=\frac{\partial n}{\partial y} \tag{14}$$

and[1]

$$\frac{d}{ds}\left(n\frac{dz}{ds}\right)=\frac{\partial n}{\partial z} \tag{15}$$

The above three equations can be combined into the following vector form

$$\frac{d}{ds}\left(n\frac{dr}{ds}\right)=\nabla n \tag{16}$$

which is known as the ray equation. The above equation can also be derived from Maxwell's equations [2], however, the algebra is much more involved.

---

[1] Equation (15) follows from the fact that one could equally well have written $ds=\sqrt{1+\dot{y}^2+\dot{z}^2}\,dx$ with dots now representing differentiation with respect to the x coordinate. Alternatively, Eq.(15) can be derived by using Eqs.(13) and (14).

The ray invariant $\tilde{\beta}$ for a waveguide

We next consider an optical waveguide with a *z* independent refractive index profile given by

$$n = n(x, y) \tag{17}$$

Thus, Eq.(15) becomes

$$\frac{d}{ds}\left(n\frac{dz}{ds}\right) = \frac{\partial n}{\partial z} = 0 \tag{18}$$

implying

$$\left(n\frac{dz}{ds}\right) = \tilde{\beta} \text{ (an invariant of the ray path)} \tag{19}$$

If θ is the angle that the ray makes with the *z* axis (Fig. 3.1) then

$$\frac{dz}{ds} = \cos\theta$$

and

$$n(x, y)\cos\theta(x, y) = \tilde{\beta} \tag{20}$$

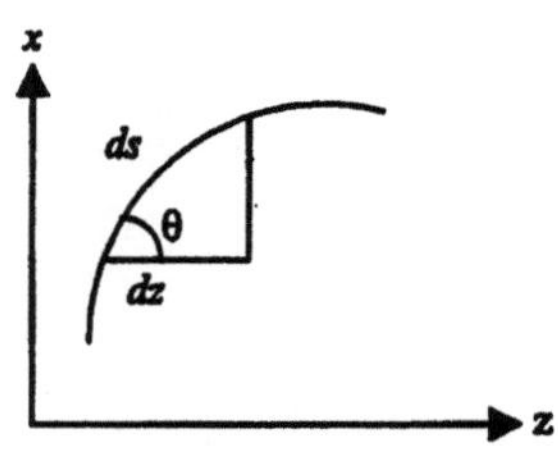

Figure 3.1 An arc length along the ray path.

The above equation implies that as the ray propagates through the waveguide, it would bend in such a way that the product n cos θ would remain unchanged. We rewrite Eq.(19) and use Eq.(4) to get

$$\tilde{\beta} = n\frac{dz}{ds} = \frac{n(x,y)}{\sqrt{1+\dot{x}^2+\dot{y}^2}} \tag{21}$$

If we use the above equation in Eq.(10) we would get

$$\frac{d}{dz}\left(\tilde{\beta}\frac{dx}{dz}\right) = \frac{n(x,y)}{\tilde{\beta}}\frac{\partial n}{\partial x} \tag{22}$$

Thus

$$\frac{d^2x}{dz^2} = \frac{1}{2\tilde{\beta}^2}\frac{\partial n^2}{\partial x} \tag{22}$$

Similarly,

$$\frac{d^2y}{dz^2} = \frac{1}{2\tilde{\beta}^2}\frac{\partial n^2}{\partial y} \tag{23}$$

The above equations represent rigorously correct ray equations for media with n^2 independent of the z coordinate.

Exact solutions for a parabolic index fiber

We consider a parabolic index fiber characterized by the following refractive index variation

$$n^2 = \begin{cases} n_1^2\left(1-2\Delta\dfrac{r^2}{a^2}\right) = n_1^2\left(1-2\Delta\dfrac{x^2+y^2}{a^2}\right) & 0<r<a \ \text{(core)} \\ n_2^2 = n_1^2(1-2\Delta) & r>a \ \text{(cladding)} \end{cases} \tag{24}$$

where n_1 and n_2 represent the core and cladding refractive indices respectively, a being the core radius and $r = a$ represents the core-cladding interface. In the core of the fiber, the refractive index is given by Eq.(24) and Eq.(22) takes the form

$$\frac{d^2x}{dz^2}+\Gamma^2 x(z)=0 \tag{25}$$

where

$$\Gamma=\frac{n_1\sqrt{2\Delta}}{a\tilde{\beta}} \tag{26}$$

Thus

$$x(z)=A\sin\Gamma z+B\cos\Gamma z \tag{27}$$

and similarly

$$y(z)=C\sin\Gamma z+D\cos\Gamma z \tag{28}$$

which represent the rigorously correct ray paths inside the core of a parabolic index fiber. The constants A, B, C and D are determined from the initial launching conditions of the ray; this is indeed very similar to the 2-dimensional harmonic oscillator problem where the trajectory of the oscillator is determined from the initial conditions. We consider below some special launching conditions:

Ray launched in the *x*-*z* plane (an example of a meridional ray)

We first consider a ray launched in the *x*-*z* plane on the *z*-axis making an angle θ_1 with the *z*-axis. Thus the launching conditions on the plane $z = 0$ are

$$y(z=0) = 0 \Rightarrow D = 0$$

$$x(z=0) = 0 \Rightarrow B = 0$$

$$\left.\frac{dy}{dz}\right|_{z=0} = 0 \Rightarrow C = 0$$

$$\left.\frac{dx}{dz}\right|_{z=0} = \tan\theta_1$$

The last equation implying

$$A = \frac{1}{\Gamma}\tan\theta_1 = \frac{a\tilde{\beta}}{n_1\sqrt{2\Delta}}\tan\theta_1$$

or

$$A = \frac{a\sin\theta_1}{\sqrt{2\Delta}} \tag{29}$$

where in the last step we have used the fact that $\tilde{\beta} = n_1\cos\theta_1$ (see Eq.(20)). Thus the ray paths are

$$x(z) = \frac{a \sin \theta_1}{\sqrt{2\Delta}} \sin\left(\frac{\sqrt{2\Delta}}{a \cos \theta_1} z\right) \tag{30}$$

$$y(z) = 0$$

This set of equations is a rigorously correct solution for a meridional ray path in a parabolic index fiber.

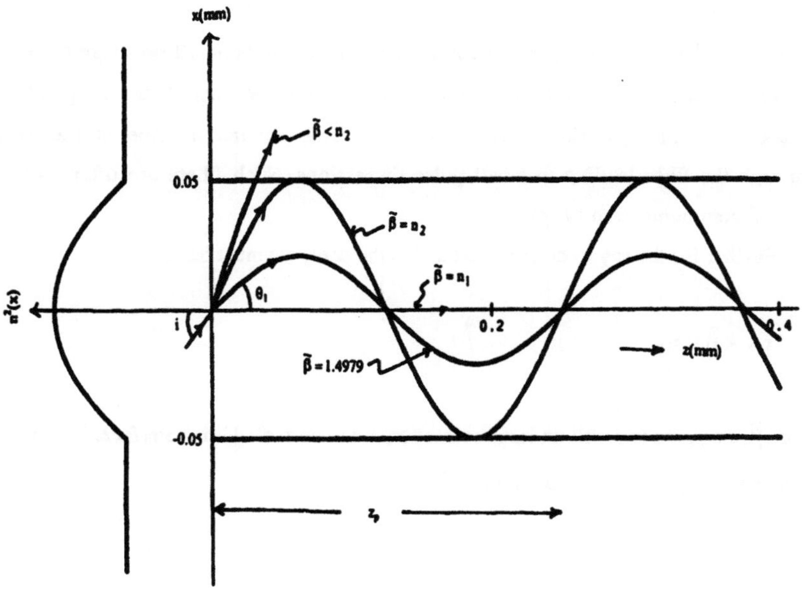

Fig. 3.2 Meridional rays in a parabolic index fiber.

The above equations describe the meridional rays confined in the *x-z* plane (Fig.3.2); we may mention that the meridional rays are defined such that they are confined in a plane and intersect the *z*-axis. Obviously, there would be meridional rays confined in the *y-z* plane---indeed in any plane containing the *z*-axis. The periodic length is given by

$$z_p = \frac{2\pi a \cos \theta_1}{\sqrt{2\Delta}} \tag{31}$$

In the paraxial approximation

$$\cos\theta_1 \approx 1$$

and

$$z_p = \frac{2\pi a}{\sqrt{2\Delta}} \tag{32}$$

which is independent of θ_1 and therefore in this approximation all rays take the same amount of time to propagate through a given length of the fiber. Thus in a parabolic index fiber, all rays take approximately the same amount of time in traversing through the fiber leading to small pulse dispersions. Such fibers are often used in optical communication systems [3].

Further for the ray to be confined inside the core, we must have

$$n_2 < \tilde{\beta} < n_1$$

For $\tilde{\beta} < n_2$, the ray will undergo refraction at the core cladding interface. This result is also valid for non-meridional rays.

Helical ray

In general, Eqs.(27) and (28) describe what are known as skew rays which do not remain confined to a plane. We consider an extreme form of skew rays in which the ray is launched on the *x*-axis (at $x = a'$) in the *y-z* plane (making angle θ' with the *z*-axis).Thus

$$x\,|_{z=0} = a' \Rightarrow B = a'$$

and

$$\left.\frac{dx}{dz}\right|_{z=0} = 0 \Rightarrow A = 0$$

Thus

$$x(z) = a' \cos \Gamma z \tag{33}$$

Further,

$$y|_{z=0} = 0 \Rightarrow D = 0$$

and

$$\left.\frac{dy}{dz}\right|_{z=0} = \tan\theta' \Rightarrow C = \frac{\tan\theta'}{\Gamma} = \frac{a\tan\theta'}{n_1\sqrt{2\Delta}}\tilde{\beta}$$

If at the launching point, $n = n'$ then $\tilde{\beta} = n'\cos\theta'$ and

$$C = \frac{an'\sin\theta'}{n_1\sqrt{2\Delta}}$$

Thus, the ray path is given by

$$\begin{aligned} x(z) &= a'\cos\Gamma z \\ y(z) &= \frac{an'\sin\theta'}{n_1\sqrt{2\Delta}}\sin\Gamma z \end{aligned} \tag{34}$$

If

$$\sin\theta' = \frac{a'n_1\sqrt{2\Delta}}{an'} = \frac{\sqrt{n_1^2 - n'^2}}{n'} \tag{35}$$

then

$$x(z) = a' \cos \Gamma z$$
$$y(z) = a' \sin \Gamma z \tag{36}$$

and

$$x^2(z) + y^2(z) = a'^2 \quad \text{independent of } z \tag{37}$$

i.e., the ray spirals around the *z*-axis as a helix at a constant distance *a'* from it (Fig.3.3). Such a ray is an extreme form of a skew ray and is known as a helical ray.

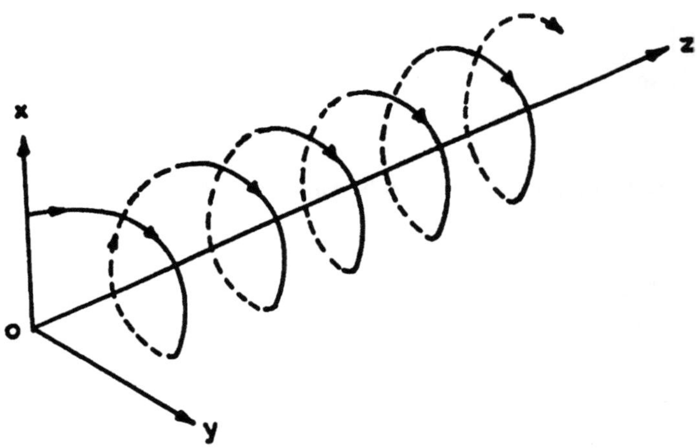

Fig. 3.3 Propagation of a helical ray in a parabolic index fiber.

The linear refractive index profile

In many problems of interest, the refractive index variation can be approximated by a linear variation

$$n^2(x) = \begin{cases} n_0^2 - \alpha x & x > 0 \\ n_0{}^2 & x < 0 \end{cases} \tag{38}$$

Then for $x > 0$

$$\frac{d^2x}{dz^2} = -\frac{\alpha}{2\tilde{\beta}^2} \tag{39}$$

Thus for a ray passing through the point $x = 0,\ z = 0$ (Fig. 3.4)

$$x(z) = -\frac{\alpha}{4\tilde{\beta}^2} z^2 + z \tan\theta_0 \tag{40}$$

with

$$\tilde{\beta} = n_0 \cos\theta_0 \tag{41}$$

Such ray paths are of interest in radio wave propagation through the ionosphere.

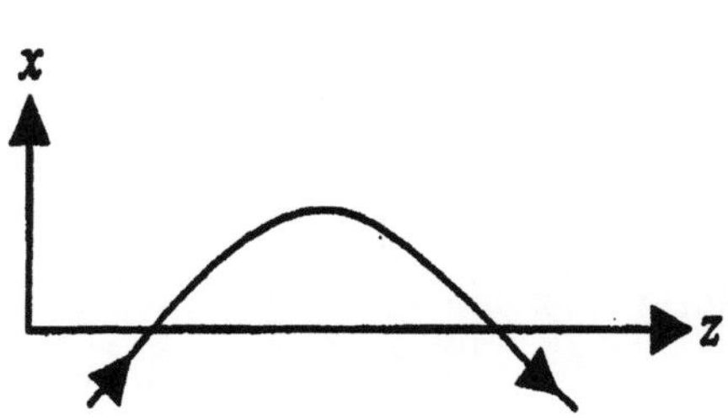

Fig. 3.4 For a linear variation of $n^2(x)$, the trajectory is a parabola.

The "sech" profile

In this example we will obtain ray paths for the following refractive index profile

$$n^2(x) = n_0^2 \operatorname{sech}^2 \alpha x \tag{42}$$

We multiply Eq.(22) by $2\frac{dx}{dz}$ and integrate to obtain

$$\left(\frac{dx}{dz}\right)^2 = \frac{n^2(x)}{n^2(x_0)\cos^2\gamma_0} - 1 \tag{43}$$

where we have assumed that at $z = 0$, $x = x_0$, $y = 0$, and $\frac{dx}{dz} = \tan\gamma_0$. Substituting for $n^2(x)$ from Eq.(42) in Eq.(43) we obtain

$$\frac{1}{\alpha}\int \frac{d\Phi}{(1-\Phi^2)^{1/2}} = \int dz \tag{44}$$

or

$$\frac{1}{\alpha}\sin^{-1}(\Phi) = z + C$$

where

$$\Phi = \frac{1}{\sqrt{A^2-1}}\sinh(\alpha x) \tag{45}$$

$$A = \frac{n_0}{n(x_0)\cos\gamma_0} \tag{46}$$

and *C* is a constant of integration. Thus the ray path is

$$x(z) = \frac{1}{\alpha}\sinh^{-1}\left\{\sqrt{A^2 - 1}\,\sin[\alpha(z + C)]\right\} \tag{47}$$

Typical ray paths are shown in Fig. 3.5. It is immediately evident that the rays are periodic in z with a period $2\pi/\alpha$ that is indeed independent of x_0 and γ_0 and hence *all* rays take the same amount of time [see chapter 4 of Ref.3].

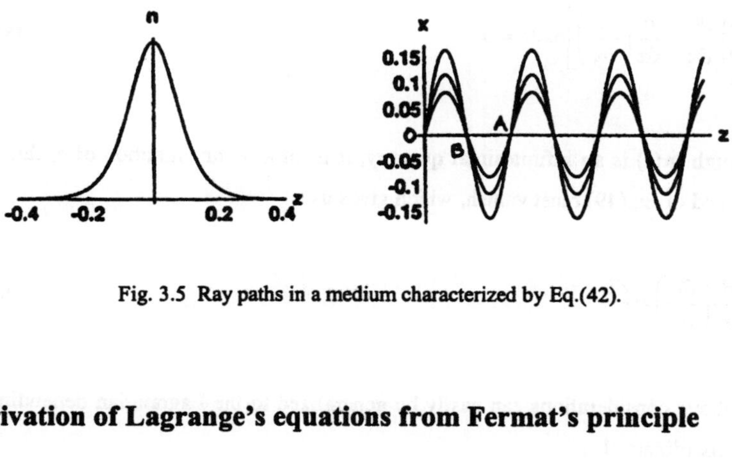

Fig. 3.5 Ray paths in a medium characterized by Eq.(42).

Derivation of Lagrange's equations from Fermat's principle

We only consider the case in which we assume the Lagrangian to be independent of the y coordinate. Fermat's principle tells us that

$$\delta \int_{z_1}^{z_2} L[x(z), \dot{x}(z), z]dz = \int_{z_1}^{z_2} \delta L[x(z), \dot{x}(z), z]dz = \int_{z_1}^{z_2}\left[\frac{\partial L}{\partial x}\delta x + \frac{\partial L}{\partial \dot{x}}\delta\dot{x}\right]dz = 0 \tag{48}$$

where δL represents the change in the Lagrangian as we go from the actual ray path to a nearby path having the same endpoints (i.e., $\delta x|_{z_1} = \delta x|_{z_2} = 0$). Now

$$\int_{z_1}^{z_2} \frac{\partial L}{\partial \dot{x}} \delta \dot{x} dz = \int_{z_1}^{z_2} \frac{\partial L}{\partial \dot{x}} \frac{d}{dz} \delta x dz$$

$$= \frac{\partial L}{\partial \dot{x}} \delta x \bigg|_{z_1}^{z_2} - \int_{z_1}^{z_2} \frac{d}{dz}\left(\frac{\partial L}{\partial \dot{x}}\right) \delta x dz$$

$$= \int_{z_1}^{z_2} \frac{d}{dz}\left(\frac{\partial L}{\partial \dot{x}}\right) \delta x dz$$

Thus Eq.(48) becomes

$$\int_{z_1}^{z_1} \left[\frac{\partial L}{\partial x} - \frac{d}{dz}\left(\frac{\partial L}{\partial \dot{x}}\right)\right] \delta x dz = 0 \tag{49}$$

Although $\delta x(z)$ is an infinitesimal quantity, it is an arbitrary function of z; thus, the integrand of Eq.(49) must vanish, which gives us

$$\frac{d}{dz}\left(\frac{\partial L}{\partial \dot{x}}\right) = \frac{\partial L}{\partial x} \tag{50}$$

The above considerations can easily be generalized to the Lagrangian depending on the y coordinate also.

Derivation of the z-component of the ray equation from Eqs.(13) and (14).

We first note that

$$\frac{d}{ds}\left(n \frac{dz}{ds}\right) = \frac{dn}{ds}\frac{dz}{ds} + n \frac{d^2 z}{ds^2} \tag{51}$$

Now

$$\left(\frac{dz}{ds}\right)^2 = 1 - \left(\frac{dx}{ds}\right)^2 - \left(\frac{dy}{ds}\right)^2 \tag{52}$$

or

$$\begin{aligned} n\frac{dz}{ds}\frac{d^2z}{ds^2} &= -\frac{dx}{ds}n\frac{d^2x}{ds^2} - \frac{dy}{ds}n\frac{d^2y}{ds^2} \\ &= \frac{dn}{ds}\left[\left(\frac{dx}{ds}\right)^2 + \left(\frac{dy}{ds}\right)^2\right] - \frac{\partial n}{\partial x}\frac{dx}{ds} - \frac{\partial n}{\partial y}\frac{dy}{ds} \end{aligned} \tag{53}$$

where we have used the x and y components of the ray equation. Substituting for $n\, d^2z/ds^2$ from Eq.(53) into Eq.(51) and using Eq.(53), one can obtain the z component of the ray equation, where one must use

$$\frac{dn}{ds} = \frac{\partial n}{\partial x}\frac{dx}{ds} + \frac{\partial n}{\partial y}\frac{dy}{ds} + \frac{\partial n}{\partial z}\frac{dz}{ds} \tag{54}$$

Ray acoustics

Study of sound propagation in the ocean is important for various applications such as ranging, detection of submarines or schools of fish or simply for underwater communication. One of the often used techniques to study this is to treat sound propagation in terms of rays and use ray equation to study the path of sound rays [4]. One can also introduce effects of reflection, scattering etc. in such an analysis and obtain the sound intensity levels at various locations. In this section we will consider a typical sound velocity profile and use the ray equation to obtain the sound ray path.

The speed of sound depends on various factors such as temperature, salinity, pressure etc. This dependence can many times lead to different sound velocity profiles along the depth direction. Thus for sound propagation studies, the ocean can be considered to be horizontally stratified with a velocity depending upon the depth. Figure 3.6 shows some typical sound velocity profiles which could have a positive or a negative gradient along the depth or could even form a sound channel much like an optical waveguide.

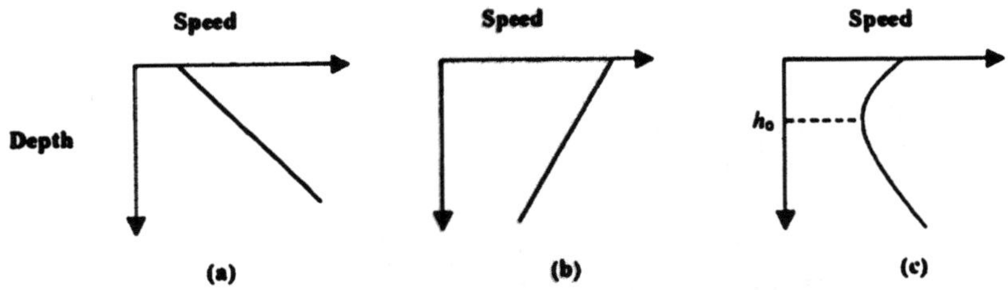

Fig. 3.6 Typical sound speed profiles along depth in a horizontally stratified ocean. In (a) the speed increases with depth, in (b) the speed decreases with depth and in (c) the speed decreases up to a certain depth and then increases.

Many times, it is convenient to approximate the sound speed variation in terms of linear segments. Thus in a certain range of depth, the speed of sound can be written as

$$c(x) = c_0 + gx \tag{55}$$

where c_0 is the sound speed at $x = 0$, x represents depth and g represents the gradient. Positive g corresponds to a positive gradient (Fig. 3.6(a)) and negative g to a negative gradient (Fig.3.6(b)). Note that the profile in Fig.3.6(c) has a negative gradient up to a certain depth and then a positive gradient beyond this. Such a velocity profile acts like a waveguide guiding sound rays to very long distances. We will now show that the ray paths in a medium characterized by a velocity profile given by Eq.(55) are circular.

For a velocity profile given by Eq.(55), we can define a refractive index for sound propagation (with respect to sound velocity at $x = 0$)

$$\begin{aligned} n(x) &= \frac{c_0}{c(x)} = \frac{c_0}{c_0 + gx} \\ &= \frac{1}{1+\tilde{g}x} \end{aligned} \tag{56}$$

where

$$\tilde{g} = \frac{g}{c_0} \tag{57}$$

Substituting in the ray equation [Eq.(43)] we get

$$\begin{aligned}\left(\frac{dx}{dz}\right)^2 &= \frac{n^2(x)}{\tilde{\beta}^2} - 1 \\ &= \frac{1}{\tilde{\beta}^2(1+\tilde{g}x)^2} - 1\end{aligned} \tag{58}$$

where $\tilde{\beta}$ is the ray invariant given by

$$\tilde{\beta} = n(x_0)\cos\theta_0 = n(x)\cos\theta(x) \tag{59}$$

where $\theta(x)$ is the angle made by the ray with the horizontal when it is at a depth x.

To show that Eq.(58) indeed represents an equation of a circle, we first consider a circle of radius R centered at (x_0, z_0). The equation of such a circle will be

$$(x-x_0)^2 + (z-z_0)^2 = R^2$$

Differentiating the above equation with respect to z, we get

$$2(x-x_0)\frac{dx}{dz} + 2(z-z_0) = 0$$

or

$$\left(\frac{dx}{dz}\right) = -\frac{z-z_0}{x-x_0}$$

Thus

$$\left(\frac{dx}{dz}\right)^2 = \frac{(z-z_0)^2}{(x-x_0)^2} = \frac{(z-z_0)^2}{(x-x_0)^2} + 1 - 1 = \frac{R^2}{(x-x_0)^2} - 1 \qquad (60)$$

This equation is of the same form as Eq.(58). Hence Eq.(58) implies that the ray path is circular. Comparing Eqs.(58) and (60), we get the radius of the circular path as

$$R = \frac{1}{\tilde{\beta}\tilde{g}} = \frac{c_0}{\tilde{\beta}g}$$

For a profile shown in Fig. 3.6(a), the ray paths will bend upward which for the case of Fig. 3.6(b), the ray paths will bend downward. (Fig. 3.7).

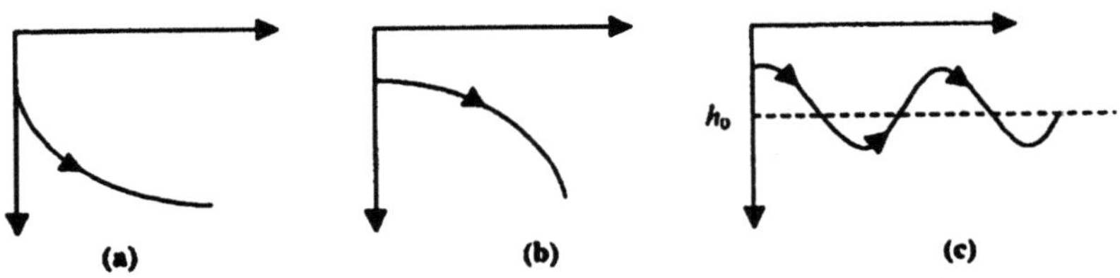

Fig. 3.7 Typical ray paths corresponding to the velocity profiles shown in Fig. 3.6. Note that in (c) the sound ray gets trapped in the ocean. Such a system is called an underwater sound channel and can be used for long distance under water communication with sound.

As a typical example we have

$$c_0 = 1500\text{m/s}$$

$$g = -0.05\text{s}^{-1}$$

If the ray is launched horizontally ($\theta = 0$) from the point $x = 0$, then

$$\tilde{\beta} = \frac{\cos\theta(x)}{1+\tilde{g}x} = 1$$

Thus

$$R = -30000\text{m}$$

The negative sign implying that the ray is curved downwards. One can indeed use this method to trace acoustic rays through any horizontally stratified water column by considering the velocity profile to consist of linear segments with different gradients.

References

1. H. Goldstein, *Classical Mechanics*, Addison-Wesley, Reading, Mass. (1960).
2. M. Born and E. Wolf, *Principles of optics*, Pergamon Press, Oxford (1975).
3. Ghatak and K. Thyagarajan, *Introduction to Fiber Optics*, Cambridge University Press (1998).
4. W.S. Burdic, *Underwater Acoustic System Analysis*, Prentice Hall, Englewood Cliffs, N.J. (1984).

4. RAY PATHS IN MEDIA WITH SPHERICAL AND CYLINDRICAL SYMMETRY

Introduction

In this chapter we will discuss the solutions of the Lagrange's equations for media having spherical and cylindrical symmetry. Ray paths in cylindrically symmetric media are of tremendous importance in fiber optics.

Media with radial symmetry

We will first show that in a medium possessing radial symmetry, i.e., where the refractive index is a function of the radial coordinate *r* alone, the rays are confined to a single plane.

Now, the ray equation [Eq.(16) of the previous chapter] can be written in the form

$$\frac{d}{ds}(n\hat{s}) = \nabla n \tag{1}$$

where $\hat{s}$ (= $d\mathbf{r}/ds$) represents the unit vector along the tangent to the ray. For a medium possessing radial symmetry, ∇n will be along $\hat{r}$. Thus

$$\hat{r}\times\frac{d}{ds}\left(n\hat{s}\right)=0 \tag{2}$$

Now,

$$\frac{d}{ds}(n\hat{s}\times\hat{r})=n\hat{s}\times\frac{d\hat{r}}{ds}+\frac{d}{ds}\left(n\hat{s}\right)\times\hat{r}=0 \tag{3}$$

Thus $n(\hat{s}\times\hat{r})$ is a constant, i.e., the rays are always confined to a plane. We choose this plane as the *x-y* plane for which $\theta=\pi/2$. Thus, along the path of the ray

$$ds=\sqrt{(dr)^2+r^2\sin^2\theta(d\theta)^2+r^2(d\phi)^2}=\sqrt{1+r^2\dot{\phi}^2}\,dr \tag{4}$$

where dots here represent differentiation with respect to *r*. Hence, the Lagrangian would be given by

$$L=n(r)\sqrt{1+r^2\dot{\phi}^2} \tag{5}$$

Now, the Lagrange equation in the variable ϕ is given by

$$\frac{d}{dr}\left(\frac{\partial L}{\partial\dot{\phi}}\right)=\frac{\partial L}{\partial\phi}=0 \tag{6}$$

Thus

$$\frac{n(r)r^2}{\sqrt{1+r^2\left(\frac{d\phi}{dr}\right)^2}}\frac{d\phi}{dr}=C \tag{7}$$

where C is a constant. The above equation would determine the ray paths.

As an example we consider the Maxwell fish eye which is characterized by the following refractive index variation

$$n(r)=\frac{n(0)}{1+(r/a)^2} \tag{8}$$

If we substitute the above form of $n(r)$ in Eq.(7) we would obtain after some rearrangement,

$$\int d\phi=\int\frac{\alpha(1+\xi^2)d\xi}{\xi\sqrt{\xi^2-\alpha^2(1+\xi^2)^2}}=\sin^{-1}\left[\frac{\alpha}{\sqrt{1-4\alpha^2}}\frac{\xi^2-1}{\xi}\right]-\beta \tag{9}$$

where $\xi = r/a$, $\alpha = C/a\, n_0$, and β is a constant of integration. Thus

$$\sin(\phi+\beta)=\sin(\phi_0+\beta)\frac{r^2-a^2}{r_0^2-a^2}\frac{r_0}{r} \tag{10}$$

which represents the path of the ray in the medium. It can be seen from Eq.(10) that this is satisfied also for $\phi = \phi_0 + \pi$ and $r = a^2/r_0$. Thus all rays emanating from (r_0, ϕ_0) intersect again at $(a^2/r_0, \phi_0 + \pi)$, and the imaging is perfect (Fig. 4.1). By transforming to Cartesian coordinates, it can easily be shown that the ray paths given by Eq.(10) are circles.

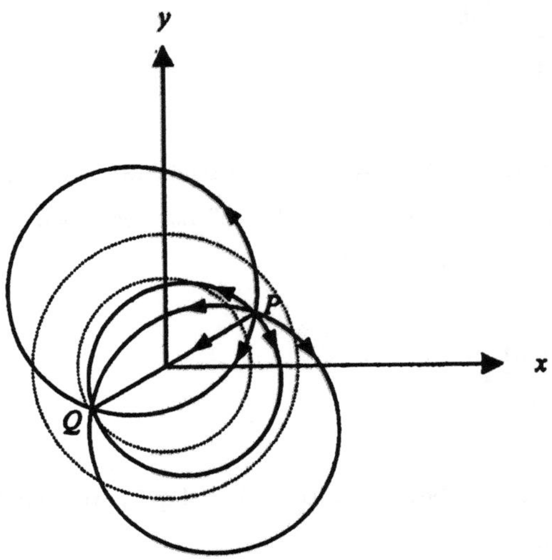

Fig. 4.1 Path of rays in Maxwell's fish eye. Loci of constant refractive index are shown as dashed circles.

Optical Lagrangian and the ray equation for cylindrically symmetric profiles*

In the previous chapter we had derived the ray equation

$$\frac{d}{ds}\left[n\frac{dr}{ds}\right] = \nabla n \tag{11}$$

where *ds* represents the arc length along the path of the ray. Now, for cylindrically symmetric media (like an optical fiber) the refractive index depends only on the cylindrical radial coordinate *r*:

$$n = n(r) \tag{12}$$

---

* The analysis in this section is based on References 1 and 2.

Thus, if we take the z component of Eq.(11) we would get

$$\frac{d}{ds}\left[n\frac{dz}{ds}\right]=\frac{\partial n}{\partial z}=0 \tag{13}$$

where we have used Eq.(12). Thus we obtain the invariant:

$$\tilde{\beta}=n\frac{dz}{ds}=n(r)\cos\theta \tag{14}$$

where θ is the angle that the ray makes with the z-axis.

Now, in a cylindrically symmetric medium (i.e., $n = n(r)$), it is obviously more convenient to use the cylindrical system of coordinates (r, ψ, z) (Fig.4.2) for which the arc length along the path of the ray is given by

$$\begin{aligned} ds &= \sqrt{(dz)^2+(dr)^2+(rd\psi)^2} \\ &= \sqrt{1+\dot{r}^2+r^2\dot{\psi}^2}\,dz \end{aligned} \tag{15}$$

where dots represent differentiation with respect to the z coordinate:

$$\begin{aligned} \dot{r} &\equiv \frac{dr}{dz} \\ \dot{\psi} &\equiv \frac{d\psi}{dz} \end{aligned} \tag{16}$$

Thus, the ray invariant $\tilde{\beta}$, is given by

$$\tilde{\beta}=n\frac{dz}{ds}=\frac{n(r)}{\sqrt{1+\dot{r}^2+r^2\dot{\psi}^2}} \tag{17}$$

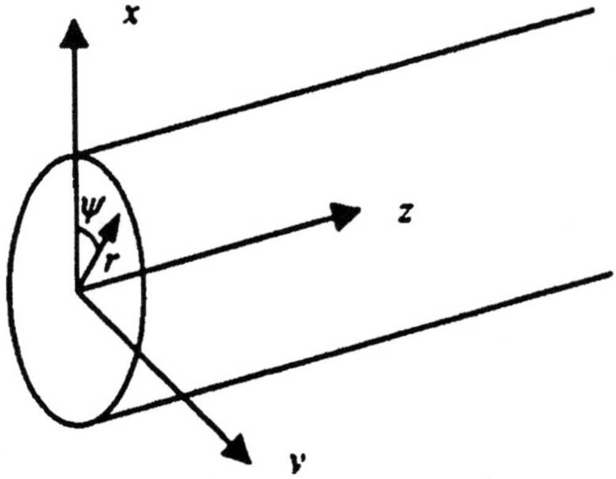

Fig. 4.2 The cylindrical system of coordinates (r, ψ, z). The z axis represents the axis of the fiber.

Furthermore, Fermat's principle gives us

$$\delta \int n ds = \delta \int n\sqrt{1+\dot{r}^2 + r^2\dot{\psi}^2}\, dz = 0 \tag{18}$$

implying

$$L = L_{opt} = n(r)\sqrt{1+\dot{r}^2 + r^2\dot{\psi}^2} \tag{19}$$

as the optical Lagrangian. The Lagrangian equation

$$\frac{d}{dz}\left(\frac{\partial L}{\partial \dot{\psi}}\right) = \frac{\partial L}{\partial \psi} \tag{20}$$

would give

$$\frac{d}{dz}\left[\frac{n(r) r^2 \dot{\psi}}{\sqrt{1+\dot{r}^2 + r^2\dot{\psi}^2}}\right] = \frac{\partial L}{\partial \psi} = 0$$

or,

$$\frac{d}{dz}\left[\tilde{\beta} r^2 \dot{\psi}\right] = 0 \tag{21}$$

where we have used Eq.(17). Thus we obtain another invariant of the ray path

$$\tilde{\beta}\frac{r^2}{a}\frac{d\psi}{dz} = \tilde{l} \tag{22}$$

In the above equation the parameter a represents the core radius which has been introduced so that the invariant $\tilde{l}$ is dimensionless. Physically, the invariant $\tilde{\beta}$ and $\tilde{l}$ are manifestations of the translational and rotational invariance of the refractive index profile. The parameter $\tilde{l}$ is usually referred to as the *skewness parameter*; obviously, for meridional rays the angle ψ remains unchanged and $\tilde{l} = 0$.

Now, from Eq.(17) we have

$$\begin{aligned}\left[\frac{n(r)}{\tilde{\beta}}\right]^2 &= 1 + \dot{r}^2 + r^2\dot{\psi}^2 \\ &= 1 + \dot{r}^2 + r^2\left(\frac{\tilde{l}a}{\tilde{\beta}r^2}\right)^2\end{aligned} \tag{23}$$

Simple rearrangement gives us

$$\dot{r} = \frac{dr}{dz} = \pm\frac{1}{\tilde{\beta}}\sqrt{f(r)} \tag{24}$$

where

$$f(r) \equiv n^2(r) - \frac{\tilde{l}^2}{(r/a)^2} - \tilde{\beta}^2 \tag{25}$$

The evaluation of the integral

$$\tilde{\beta} \int \frac{dr}{\sqrt{f(r)}} = \pm \int dz \tag{26}$$

would give us the r coordinate of the ray as a function of z as it propagates through the fiber. Once $r(z)$ is known, the integration of the equation

$$\dot{\psi} = \frac{d\psi}{dz} = \frac{a\tilde{l}}{\tilde{\beta} r^2(z)} \tag{27}$$

would determine the azimuthal co-ordinate ψ as a function of z. Obviously, $f(r)$ should be positive; this condition will be used later for classification of different types of rays.

Exact ray paths in a parabolic-index fiber

In Sec.3.4, we had obtained ray paths in a parabolic index fiber using the Cartesian system of coordinates. We will now use Eq.(24) to determine the exact ray paths in a parabolic index fiber for which the refractive index variation (inside the core) is given by

$$n^2(r) = n_1^2 \left[1 - 2\Delta\left(\frac{r}{a}\right)^2\right] = n_1^2 - g^2 r^2 \tag{28}$$

where

$$g^2 = n_1^2 \frac{2\Delta}{a^2} \tag{29}$$

Typical parameters for a parabolic index fiber are

$n_1 \approx 1.45$

$\Delta \approx 0.01$

$a \approx 25\mu m$

implying

$g \approx 8.2 \times 10^3 \, m^{-1}$

Substituting the parabolic refractive index variation in Eq.(25), Eq.(26) takes the form

$$\tilde{\beta} \int \frac{dr}{\sqrt{n_1^2 - g^2 r^2 - \dfrac{\tilde{l}^2}{(r/a)^2} - \tilde{\beta}^2}} = \pm \int dz \tag{30}$$

Carrying out the integration we obtain

$$r^2(z) = \gamma + \alpha \cos\left(\frac{2g}{\tilde{\beta}}(z - z_1)\right) \tag{31}$$

which represents exact ray paths in a parabolic index fiber. In Eq.(31) z_1 is the constant of integration,

$$\gamma = \frac{n_1^2 - \tilde{\beta}^2}{2g^2} = \frac{1}{2}\left[1 - \frac{\tilde{\beta}^2}{n_1^2}\right]\frac{a^2}{2\Delta} \tag{32}$$

and

$$\alpha^2 = \gamma^2 - \frac{\tilde{l}^2 a^2}{g^2} = \left(\frac{a^2}{2\Delta}\right)^2\left[\frac{1}{4}\left(1 - \frac{\tilde{\beta}^2}{n_1^2}\right)^2 - \tilde{l}^2 \frac{2\Delta}{n_1^2}\right] \tag{33}$$

Since

$$-1 < \cos\left(\frac{2g}{\widetilde{\beta}}(z - z_1)\right) < 1$$

we have

$$\gamma - \alpha < r^2(z) < \gamma + \alpha \tag{34}$$

as shown in Fig.4.3. The circles

$$r(z) = \sqrt{\gamma - \alpha} = r_1$$

and

$$r(z) = \sqrt{\gamma + \alpha} = r_2$$

represent the inner and outer caustics where *f(r)* [and hence *dr/dz*] vanishes (see Fig. 4.3). For $\widetilde{l} = 0$, $\alpha = \gamma$ and the radius of the inner caustic becomes zero and the ray intersects the axis. In order to obtain $r(\psi)$ we divide Eq.(22) by Eq.(25) to obtain

$$\widetilde{\beta}\frac{r^2}{a}\frac{d\psi}{dr} = \frac{\widetilde{l}\widetilde{\beta}}{\sqrt{f(r)}} \tag{35}$$

or

$$\int d\psi = \pm a\widetilde{l} \int \frac{dr}{r^2\sqrt{n^2(r) - \dfrac{\widetilde{l}^2}{(r/a)^2} - \widetilde{\beta}^2}} \tag{36}$$

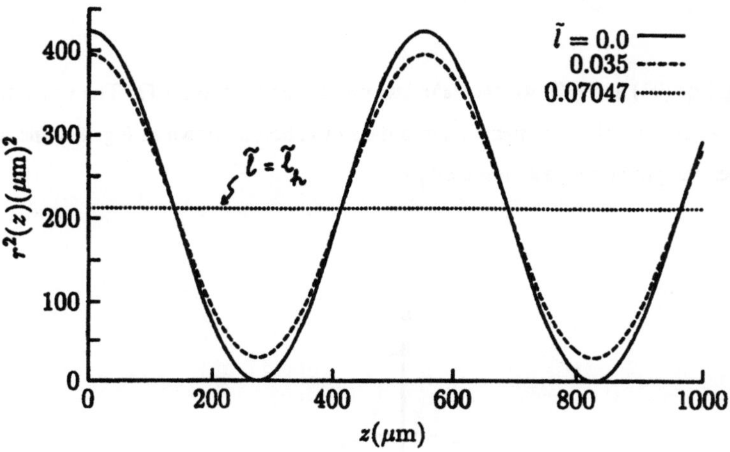

Fig. 4.3 Variation of r^2 with z for different values of $\tilde{l}$. The values of various parameters are $n_1 = 1.45, \Delta = 0.01, a = 25\ \mu m$.

If we now substitute Eq.(28) for $n^2(r)$ and carry out the integration we would get

$$\frac{1}{r^2} = \nu + \mu \cos 2(\psi - \psi_1) \tag{37}$$

where ψ_1 is a constant of integration,

$$\nu = \frac{n_1^2 - \tilde{\beta}^2}{2\tilde{l}^2 a^2} = \frac{\gamma}{\tilde{l}^2}$$
$$l = \frac{\tilde{l} a}{g} \tag{38}$$

and

$$\mu = \sqrt{\nu^2 - \frac{g^2}{\tilde{l}^2 a^2}} = \sqrt{\frac{\gamma^2}{l^4} - \frac{1}{l^2}} \tag{39}$$

Using Eqs.(31) and (37) we can calculate r and ψ as functions of z. The projection of the ray path in a plane perpendicular to the z direction is shown in Fig. 4.4 and as can be seen the projection is a closed ellipse.

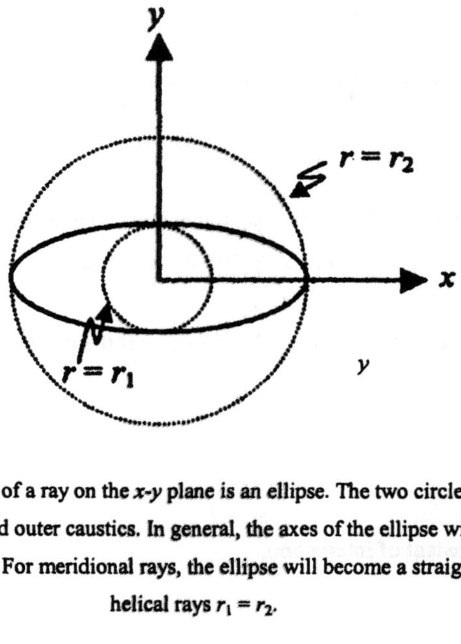

Fig. 4.4 The projection of a ray on the x-y plane is an ellipse. The two circles ($r = r_1$ and $r = r_2$) represent the inner and outer caustics. In general, the axes of the ellipse will make an angle with the x- and y-axes. For meridional rays, the ellipse will become a straight line, and for helical rays $r_1 = r_2$.

We next consider some special cases:

Meridional rays ($\tilde{l} = 0$): For meridional rays, $\tilde{l} = 0$ and

$$\alpha = \gamma = \frac{1}{2}\left(1 - \frac{\tilde{\beta}^2}{n_1^2}\right)\frac{a^2}{2\Delta}$$

Thus, Eq.(31) can be written in the form

$$\begin{aligned} r^2(z) &= \gamma\left[1+\cos\frac{2g}{\tilde{\beta}}(z-z_1)\right] \\ &= 2\gamma\cos^2\left[\frac{g}{\tilde{\beta}}(z-z_1)\right] \end{aligned} \tag{40}$$

or

$$\begin{aligned} r(z) &= \pm\sqrt{2\gamma}\cos\left[\frac{g}{\tilde{\beta}}(z-z_1)\right] \\ &= \pm\sqrt{\left(1-\frac{\tilde{\beta}^2}{n_1^2}\right)\frac{a^2}{2\Delta}}\cos\left(\frac{n_1\sqrt{2\Delta}}{a\tilde{\beta}}z-\phi_1\right) \end{aligned} \tag{41}$$

Since r should always be positive, we must choose the + sign where the cosine function is positive and - sign when the cosine function is negative. Further, when $\tilde{l}=0$, we must have

$$\cos 2(\psi-\psi_1)=0$$

or

$$(\psi-\psi_1)=\frac{\pi}{2},\frac{3\pi}{2}$$

implying that ψ should not change with the z-coordinate except for abrupt changes from $\pi/2$ to $3\pi/2$.

Helical rays ($\tilde{l}=\tilde{l}_h$): From Eq.(33) one can readily see that the maximum value of $\tilde{l}$ is given by

$$\tilde{l} = \tilde{l}_h = \frac{\gamma g}{a} = \frac{n_1^2 - \tilde{\beta}^2}{2n_1\sqrt{2\Delta}} \tag{42}$$

and $\alpha = 0$ so that

$$r^2(z) = \gamma = \frac{1}{2}\left(1 - \frac{\tilde{\beta}^2}{n_1^2}\right)\frac{a^2}{2\Delta} \tag{43}$$

implying that the ray would always be at a constant distance from the axis; this is referred to as the helical ray.

Classification of rays

We will now discuss the different types of rays that can propagate through an optical fiber. Our starting point will be the equation determining the ray path [see Eq.(24)]

$$\tilde{\beta}\int\frac{dr}{\sqrt{f(r)}} = \int dz \tag{44}$$

where

$$f(r) = n^2(r) - \tilde{\beta}^2 - \frac{\tilde{l}^2}{(r/a)^2} \tag{45}$$

Obviously, for a ray path to be allowed, $f(r)$ should be positive. The forbidden regions correspond to $f(r)$ being negative.

We will assume the refractive index variation to be of the form

$$n^2(r) = n_1^2\left[1-2\Delta\left(\frac{r}{a}\right)^q\right] \qquad 0<r<a$$
$$= n_2^2 = n_1^2(1-2\Delta) \qquad r>a \tag{46}$$

which is known as the power law profile (see Fig.5.1). We may mention that most multimode optical fibers can be approximately characterized by a power law profile. We will first consider the case $\tilde{l}=0$ which will be followed by the case $\tilde{l}>0$.

$\tilde{l}=0$ **: Meridional rays:** From Eq.(45), for $\tilde{l}=0$,we have

$$f(r) = n^2(r) - \tilde{\beta}^2 \tag{47}$$

The $f(r)$ variations for $\tilde{\beta}^2 > {n_1}^2$, ${n_2}^2 < \tilde{\beta}^2 < {n_1}^2$ and $\tilde{\beta}^2 < {n_2}^2$ are shown in Figs. 4.5(a), (b) and (c) respectively. Although the figures correspond to the parabolic index profile, the general behavior will be the same for an arbitrary power law variation. As can be seen, for $\tilde{\beta}^2 > {n_1}^2$, $f(r)$ is everywhere negative and no ray path is possible.

This is also obvious from the fact that

$$\tilde{\beta} = n(r)\cos\theta(r) \tag{48}$$

and since the maximum values of $n(r)$ and cos θ are n_1 and 1, $\tilde{\beta}$ can never be greater than n_1, i.e.,

$$\tilde{\beta} \le n_1 \tag{49}$$

When

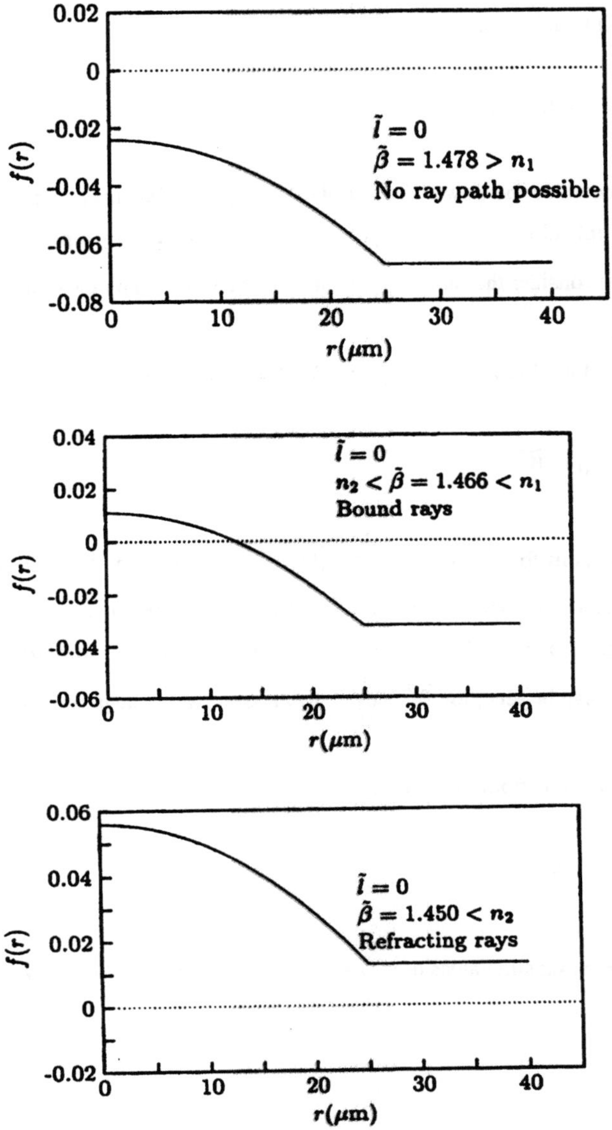

Fig. 4.5 Variation of $f(r)$ for meridional rays ($\tilde{l} = 0$). No ray paths are possible for $\tilde{\beta} > n_1$ (see (a)). Bound rays and refracting rays correspond to $n_2 < \tilde{\beta} < n_1$ and $0 < \tilde{\beta} < n_1$, respectively (see (b) and (c)).

$$n_2^2 < \tilde{\beta}^2 < n_1^2 \tag{50}$$

$f(r)$ is positive in the domain $0 < r < r_2$ [see Fig.4.4(b)] and $r = r_2$ is such that

$$n^2(r_2) = \tilde{\beta}^2 \tag{51}$$

These are the *bound meridional rays* and $r = r_2$ is known as the turning point which is given by

$$r_2 = a\left[\frac{n_1^2 - \tilde{\beta}^2}{2\Delta n_1^2}\right]^{1/q} \tag{52}$$

For a step index fiber ($q = \infty$) and $r_2 = a$; the rays undergo total internal reflection at the core cladding interface.

Finally, for $\tilde{\beta}^2 < n_2^2$, $f(r)$ is everywhere positive (see Fig.4.4(c)) and the ray just refracts away at the core cladding interface.

Skew rays: $\tilde{l} > 0$ **:** We next consider the case $\tilde{l} > 0$ which correspond to skew rays where the rays are not confined to a single plane. Once again, for $\tilde{\beta}^2 > n_1^2$, $f(r)$ will be everywhere negative implying that no ray paths are possible.

For $n_2^2 < \tilde{\beta}^2 < n_1^2$, the variations of $n^2(r) - \tilde{\beta}^2$ and $\dfrac{\tilde{l}^2}{(r/a)^2}$ for $q = 2$ are shown in Fig.4.6(a). The points of intersection of the two curves determine the domain of the ray path. When the value of $\tilde{l}$ is such that the two curves touch at one point, we have a helical ray ($\tilde{l} = \tilde{l}_h$); the value of $\tilde{l}_h$ will depend on the value of $\tilde{\beta}$ In Fig.4.6(b) we have plotted the corresponding variation of $f(r)$. Figure 4.6 corresponds to $q = 2$ with

$$\begin{aligned} n_1 &= 1.47 \\ a &= 25\mu m \\ \Delta &= 0.01 \end{aligned} \tag{53}$$

and

$$\tilde{\beta} = 1.46 \tag{54}$$

Thus from Eq.(42) we find

$$\tilde{l}_h = 0.07047 \tag{55}$$

For $\tilde{l} < \tilde{l}_h$, we will have 2 values of r (equal to r_1 and r_2) where $f(r) = 0$; the points r_1 and r_2 at which $f(r) = 0$ (inside the core) are known as the turning points and the cylindrical surfaces $r = r_1$ and $r = r_2$ are known as the inner and outer caustics respectively. For $q = 2$ the projection of a bound skew ray path on the x-y plane is an ellipse as shown in Fig.4.3.

For a step index fiber ($q = \infty$), $r_2 = a$, i.e., the outer caustic is at the core-cladding interface. Figure 4.7 corresponds to a step index fiber ($q = \infty$) with the same values of n_1, a, Δ and $\tilde{\beta}$ as given by Eqs.(53) and (54). Obviously,

$$\tilde{l}_h = \sqrt{n_1^2 - \tilde{\beta}^2} = 0.17117 \tag{56}$$

and the helical ray slides along the core cladding interface.

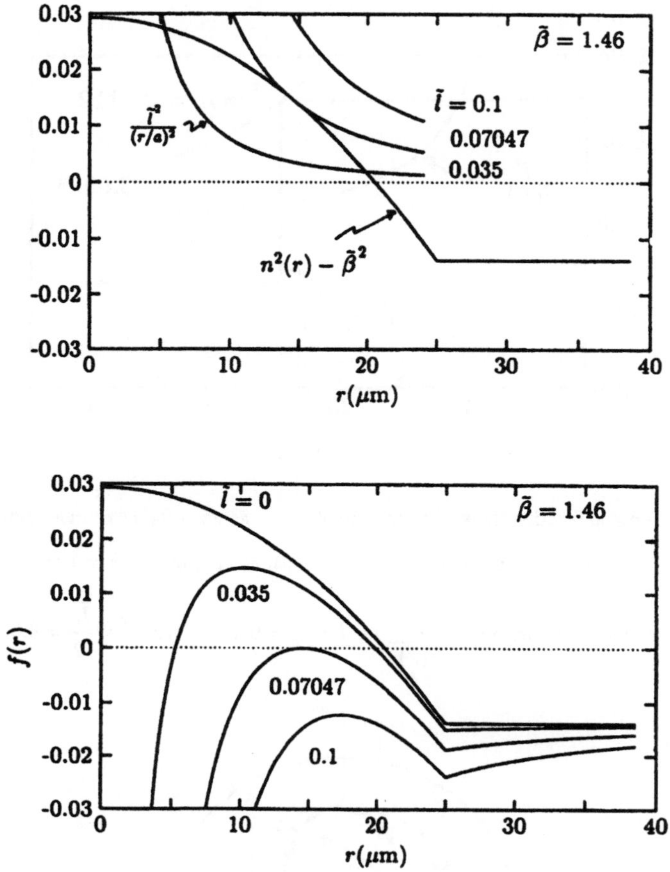

Fig. 4.6 Variations of $n^2(r) - \tilde{\beta}^2$ and $\frac{\tilde{l}^2}{(r/a)^2}$ corresponding to a $q = 2$ fiber for bound rays ($n_2^2 < \tilde{\beta}^2 < n_1^2$). (b) The corresponding $f(r)$ variations; the values of various parameters are n_1=1.47, $\Delta = 0.01$, $a = 25\mu m$, and $\tilde{\beta} = 1.46$. The value of $\tilde{l}_h = 0.07047$.

For $\tilde{l} > \tilde{l}_h$, $f(r)$ will be negative everywhere and there will be no ray path possible. For $q \neq 2$ and $\neq \infty$, the projection of the bound skew ray on the x-y plane is qualitatively shown in Fig.4.8; for meridional rays ($\tilde{l} = 0$) the curve will degenerate into a straight line.

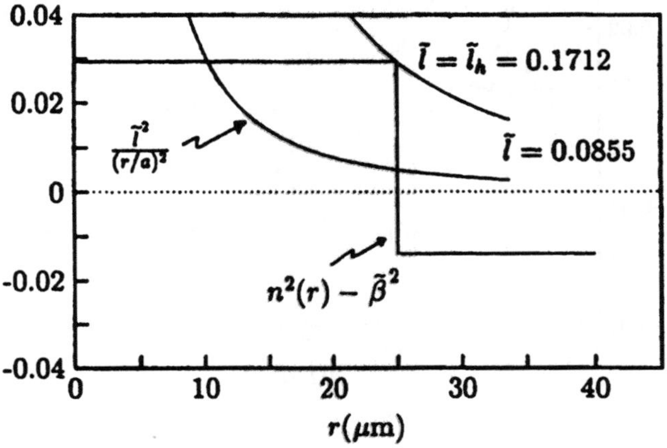

Fig. 4.7 The step function shows the variation of $n^2(r) - \tilde{\beta}^2$ and r for a step index fiber; the values of various parameters are $n_1 = 1.47, \Delta = 0.001, a = 25\ \mu m$, and $\tilde{\beta} = 1.46$. The two curves represent the variation of $\frac{\tilde{l}^2}{\left(r/a\right)^2}$ for $\tilde{l} = \tilde{l}_h = 0.17117$ and $\tilde{l} = \frac{\tilde{l}_h}{2} = 0.0855$.

Fig. 4.8 Projection of the ray path onto the fiber cross section for $q \neq 2$, $\tilde{l} \neq 0$. For $\tilde{l} = 0$, the ray is meridional with $r_1 = 0$ and the projection is a straight line. For a parabolic index fiber ($q = 2$), the projection is a closed ellipse. [Adapted from Ref. 1]

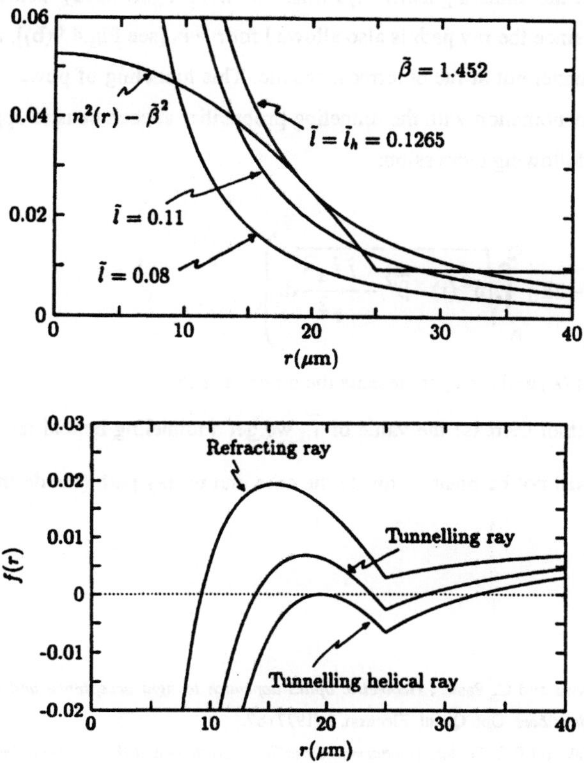

Fig. 4.9 (a) For $\tilde\beta(=1.452) < n_2$, variations of $n^2(r) - \tilde\beta^2$ and $\frac{\tilde l^2}{(r/a)^2}$ for $\tilde l = \tilde l_h = 0.1265$ (tunneling helical ray), $\tilde l = 0.11$ (tunneling skew ray), and $\tilde l = 0.08$ (refracting ray) in a parabolic index fiber ($n_1 = 1.47, \Delta = 0.01, a = 25\ \mu m$). (b) The corresponding variations of $f(r)$ $[= n^2(r) - \tilde\beta^2 - \frac{\tilde l^2}{(r/a)^2}]$.

We finally consider the case when $\tilde\beta^2 < n_2^{\,2}$. Corresponding to different values of $\tilde l$, four different situations may arise. For small values of $\tilde l$, there is only one point of intersection between the two curves and we have a refracting (skew) ray as shown in Fig.4.9(a). As we increase the value of $\tilde l$, we obtain three values of r where $f(r)$ =

0 and we have the tunneling leaky rays where we have a guided ray inside the core of the fiber and since the ray path is also allowed for $r > r_3$ [see Fig.4.9(b)], a part of the energy will tunnel out at the outermost caustic. This tunneling of power is entirely a wave optic phenomenon with the tunneling probability at each caustic approximately given by the following expression:

$$T \cong \exp\left(-2k_0 \int_{r_1}^{r_2} \sqrt{n^2(r) - \tilde{\beta}^2 - \frac{\tilde{l}^2 a^2}{r^2}}\, dr\right)$$

where $k_0 = 2\pi/\lambda_0$ and $r = r_2$ represents the outer caustic.

As we further increase the value of $\tilde{l}$ we get a tunneling helical ray $\tilde{l} = \tilde{l}_h$. For $\tilde{l} > \tilde{l}_h$, $f(r)$ will not be positive inside the core and no ray path (inside the core) will be possible.

References

1. A. Ankiewicz and C. Pask, , *Geometric optics approach to light acceptance and propagation in graded index fibers*, Opt. Quant. Electron. **9** (1977) 87.
2. A.K. Ghatak and E.G. Sauter, *The harmonic oscillator problem and the parabolic index waveguide*, European J. Physics, **10** (1989) 136.

5. RAY PATHS IN BENT WAVEGUIDES

Introduction

The study of ray paths in graded-index multimode media is a subject of considerable interest in the area of fiber optics and graded-index imaging systems [1-12]. When rays in multimode waveguides encounter bends there are radiation losses; these losses are either by refraction or by tunneling. The fractional loss of power when a ray is reflected from an outer caustic along the ray path is usually calculated by using the WKB method. Hence it is essential to know the exact ray paths in bent waveguides and thereby know the exact positions of the ray caustics to calculate the bend loss. In this chapter, we present an analysis to find ray paths in bent slab waveguides.

Optical Lagrangian in cylindrical coordinates

Although we had discussed the Lagrange's equations in cylindrical coordinates in Chapter 4, we briefly repeat some of the equations here as there is slight change in notation. We start with the Fermat's principle

$$\delta \int n ds = 0 \tag{1}$$

where

$$n = n(r, \phi, z) \tag{2}$$

where (r, ϕ, z) represent the cylindrical system of coordinates (Fig. 5.1). The arc length ds along the path of the ray can be written in the form

$$ds = \sqrt{1 + (r\dot{\phi})^2 + (\dot{z})^2}\, dr \tag{3}$$

where

$$\dot{\phi} = \frac{d\phi}{dr} \qquad \dot{z} = \frac{dz}{dr} \tag{4}$$

Thus Eq.(1) can be written in the form

$$\delta \int L dr = 0$$

where

$$L = n(r, \phi, z)\sqrt{1 + (r\dot{\phi})^2 + (\dot{z})^2} \tag{5}$$

represents the optical Lagrangian in the cylindrical system of coordinates. In the following sections, we use the Lagrangian given by Eq.(5) to determine the ray paths for specific types of refractive-index profile which are of considerable practical importance.

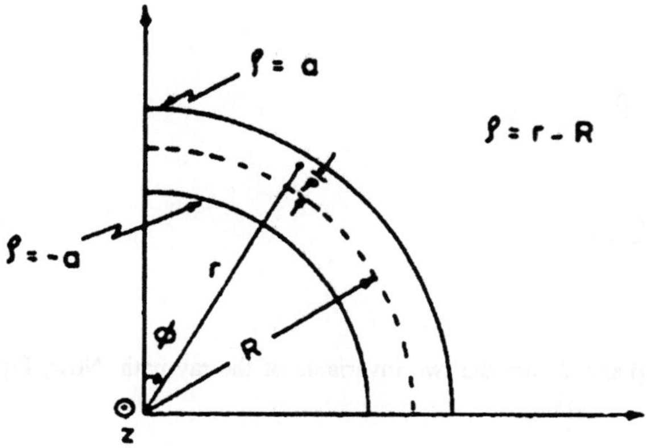

Fig. 5.1 The cylindrical system of coordinates and a slab waveguide bent along the arc of a circle of radius *R*.

Ray paths in cylindrically symmetric (*z*-independent) refractive-index profiles

We assume that the refractive-index depends only on the *r*-coordinate, i.e.,

$$n = n(r) \tag{6}$$

This is indeed the case for a straight optical fiber. Even for a bent slab waveguide, the profile is essentially of the form given by Eq.(6), provided we choose our coordinate system appropriately (see Sec. 5.5). Thus

$$L = n(r)\sqrt{1+(r\dot{\phi})^2+(\dot{z})^2}$$

and the Lagrange's equations

$$\frac{d}{dr}\left[\frac{\partial L}{\partial \dot{z}}\right] = \frac{\partial L}{\partial z} = 0$$

$$\frac{d}{dr}\left[\frac{\partial L}{\partial \dot{\phi}}\right] = \frac{\partial L}{\partial \phi} = 0$$

readily give the following invariants:

$$\frac{\partial L}{\partial \dot{z}} = n\frac{dz}{ds} = \tilde{\beta} \tag{7}$$

$$\frac{\partial L}{\partial \dot{\phi}} = nr^2\frac{d\phi}{ds} = \tilde{l} \tag{8}$$

The quantities $\tilde{\beta}$ and $\tilde{l}$ are the two invariants of the ray path. Now, Eq.(3) can be written as

$$\frac{dr}{ds} = \pm\sqrt{1-\left(r\frac{d\phi}{ds}\right)^2-\left(\frac{dz}{ds}\right)^2}$$

Substituting for the terms on the right-hand side from Eqs.(7) and (8) we obtain

$$\frac{dr}{dz} = \pm\frac{1}{\tilde{\beta}}\sqrt{f(r)} \tag{9}$$

where

$$f(r) = n^2(r) - \tilde{\beta}^2 - \frac{\tilde{l}^2}{r^2} \tag{10}$$

The above equations are similar to the ones derived in Chapter 4.

Ray invariants and ray equations for profiles of the form $n^2(r) = n_1{}^2(r) + n_2{}^2(z)$

We next consider a special class of cylindrically symmetric refractive-index profiles which can be put in a separable form, i.e.,

$$n^2(r, z) = n_1^2(r) + n_2^2(z) \tag{11}$$

In this section, we show that for a separable profile given by the above equation, it is possible to derive equations, simple integrations of which would give the exact ray paths. In the next section, we show that the profile given by Eq.(11) is applicable to bent waveguides with a separable profile; i.e., the (unbent) waveguide is characterized by a refractive-index distribution of the form

$$n^2(x, y) = n'^2(x) + n''^2(y) \tag{12}$$

Obviously, a parabolic-index fiber has a separable profile:

$$\begin{aligned} n^2(r) &= n_0^2\left[1 - 2\Delta(r/a)^2\right] \\ &= n_0^2 - gx^2 - gy^2 \end{aligned}$$

where $g = 2\Delta(n_0{}^2/a^2)$, and a represents the radius of the core. If the axis of such a fiber is bent along the arc of a circle, we will have a profile of the type given by Eq.(11). Furthermore, it has been shown that a separable profile given by Eq.(12) can describe a large class of optical waveguides (including rectangular core integrated optical waveguides and elliptical core fibers) to a considerable degree of accuracy.

For $n^2(r, z)$ given by Eq.(11), the Lagrange's equation

$$\frac{d}{dr}\left(\frac{\partial L}{\partial \dot{z}}\right) = \frac{\partial L}{dz}$$

gives us

$$\frac{d}{dr}\left(\frac{n\dot{z}}{\sqrt{1+r^2\dot{\phi}^2+\dot{z}^2}}\right)=\frac{\sqrt{1+r^2\dot{\phi}^2+\dot{z}^2}}{2n(r,z)}\frac{dn_2^2(z)}{dz}$$

which can be written in the form

$$2\left[n(r,z)\frac{dz}{ds}\right]\frac{d}{dz}\left[n(r,z)\frac{dz}{ds}\right]=\frac{dn_2^2(z)}{dz}$$

Defining

$$G=n(r,z)\frac{dz}{ds} \tag{13}$$

the above equation gives

$$\frac{d}{dz}\left[G^2-n_2^2(z)\right]=0$$

Thus we obtain a ray invariant $\tilde{l}$ as

$$G^2(z)-n_2^2(z)=\left(n(r,z)\frac{dz}{ds}\right)^2-n_2^2(z)=\tilde{l}^2 \tag{14}$$

The Lagrange's equation corresponding to the ϕ-coordinate gives us

$$\frac{d}{dr}\left(n(r,z)r^2\frac{d\phi}{ds}\right)=\frac{\partial L}{\partial\phi}=0$$

and we get the second ray invariant:

$$n(r,z)r^2\frac{d\phi}{ds}=\tilde{\gamma} \tag{15}$$

The equation defining the exact ray paths can now be obtained. We introduce a new invariant $\tilde{\beta}$ as

$$\tilde{\beta}=\frac{\tilde{\gamma}}{R}=n(r,z)\frac{r^2}{R}\frac{d\phi}{ds} \tag{16}$$

Now, using Eq.(14),

$$[\tilde{l}^2+n_2^2(z)]\frac{1}{n^2(r,z)}=\left(\frac{dz}{ds}\right)^2=\left(\frac{dz}{d\phi}\frac{d\phi}{ds}\right)^2$$

Substituting for $d\phi/ds$ from Eq.(16) in the above equation we obtain

$$\frac{dz}{d\phi}=\pm\frac{r^2(\phi)}{\tilde{\beta}R}\sqrt{F(z)} \tag{17}$$

where

$$F(z)=\sqrt{n_2^2(z)+\tilde{l}^2} \tag{18}$$

and $r(\phi)$ is determined from Eq.(20). Now, using Eq.(16) we may write

$$\left(\frac{n(r,z)r^2}{\tilde{\beta}R}\right)^2=\left(\frac{ds}{d\phi}\right)^2=\left(\frac{dr}{d\phi}\right)^2+r^2+\left(\frac{dz}{d\phi}\right)^2 \tag{19}$$

Substituting for $dz/d\phi$ from Eq.(17) we get

$$\frac{dr}{d\phi} = \pm \frac{r}{\tilde{\beta}} \sqrt{\frac{\left(n_1^2(r) - \tilde{l}^2\right) r^2}{R^2} - \tilde{\beta}^2} \tag{20}$$

Equations (17) and (20) represent (for a separable profile) the rigorously correct ray equations which can be used to determine the ray paths in bent waveguides.

Ray paths in a bent slab waveguide

We consider a slab waveguide characterized by the following refractive-index distribution:

$$n^2 = n^2(x)$$

We assume such a waveguide to be bent along the arc of a circle of radius R as shown in Fig. 5.1. In the cylindrical system of coordinates defined in Fig. 5.1, we have

$$n^2 = n^2(r) = n_1^2(r)$$

We can use the analysis of Sec 5.4 with $n_2^2(z) = 0$ to obtain the ray invariants

$$\tilde{l} = n(r)\frac{dz}{ds} \tag{21}$$

$$\tilde{\beta} = n(r)\frac{r^2}{R}\frac{d\phi}{ds} \tag{22}$$

and the ray equations

$$\frac{dz}{d\phi} = \pm \frac{r^2 \tilde{l}}{\tilde{\beta} R} \tag{23}$$

$$\frac{dr}{d\phi} = \pm \frac{r}{\tilde{\beta}} \sqrt{\frac{\left(n_1^2(r) - \tilde{l}^2\right) r^2}{R^2} - \tilde{\beta}^2} \tag{24}$$

Introducing the variables $\rho = r - R$ and $\xi = R\,\phi$, we get

$$\frac{dz}{d\xi} = \pm \frac{1}{\tilde{\beta}} \left(1 + \frac{\rho}{R}\right)^2 \tilde{l} \tag{25}$$

$$\frac{d\rho}{d\xi} = \pm \frac{1}{\tilde{\beta}} \left(1 + \frac{\rho}{R}\right) \sqrt{f(\rho)} \tag{26}$$

where

$$f(\rho) = \left(n_1^2(\rho) - \tilde{l}^2\right)\left(1 + \frac{\rho}{R}\right)^2 - \tilde{\beta}^2 \tag{27}$$

From the initial launch conditions we can determine the values of the ray invariants $\tilde{\beta}$ and $\tilde{l}$ and solve Eqs.(25) and (26) explicitly to obtain the ray paths $\rho = \rho(\xi)$ and $z = z(\xi)$. We consider now the case of a parabolic index bent slab waveguide, i.e., $n^2 = n^2(\rho) = n_0^2 - g\rho^2$. It is easily seen from Eqs.(25) and (26) that when $\tilde{l} = 0$ the ray path is confined to a single plane, and the z-coordinate of the ray remains constant.

We first consider small launch angles and refer to curve 2 in Fig. 5.2. We see that $f(\rho)$ is positive in the regions $\rho_2 < \rho < \rho_3$ and for $\rho > \rho_4$ and negative in the region $\rho_3 < \rho < \rho_4$. This is the case of what is known as tunneling rays which undergo partial reflection at $\rho = \rho_3$ and part of the energy tunnels through the evanescent region $\rho_3 < \rho < \rho_4$ to leak away beyond $\rho = \rho_4$. If the launch angle is sufficiently large, we have a situation similar to the one shown in curve 3a where $f(\rho)$ is positive for all values of $\rho > \rho_1$, and we have what are known as refracting rays which leak out very quickly from the core. It is obvious that for a bent structure there are no bound rays, and all rays launched are either tunneling or refracting.

The exact ray paths can be obtained from Eqs. (25) and (26). However the ± sign in front of the square root is chosen according to the slope of the ray path and is hence difficult to incorporate into a numerical algorithm. This problem can be overcome by squaring and differentiating the two first-order simultaneous differential equations to obtain two second-order simultaneous differential equations given by

$$\begin{aligned}\frac{d^2\rho}{d\xi^2} &= 2\left[\frac{n_1^2(\rho)-\tilde{l}^2}{\tilde{\beta}^2}\right]\left(1+\frac{\rho}{R}\right)^3\frac{1}{R}\\ &\quad+\frac{1}{2\tilde{\beta}^2}\left(1+\frac{\rho}{R}\right)^4\frac{dn_1^2(\rho)}{d\rho}-\left(1+\frac{\rho}{R}\right)\frac{1}{R}\end{aligned} \tag{28}$$

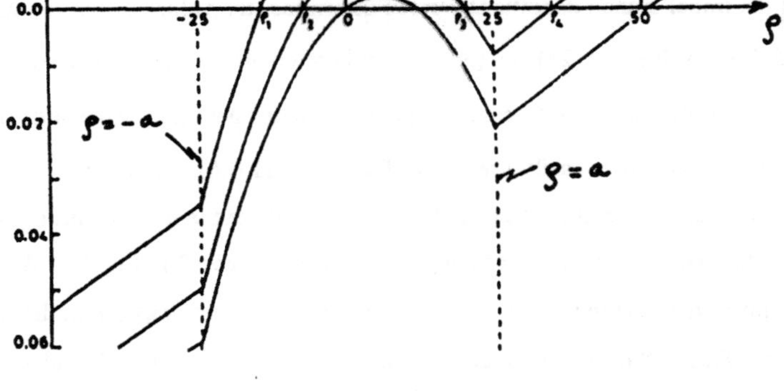

Fig. 5.2 Variation of $f(\rho)$ with ρ for a bent parabolic-index slab waveguide corresponding to $\tilde{l}=0$. Curves 1,2, and 3 correspond to initial on-axis launch angles equal to 0.01 ($\tilde{\beta}=2.204$), 0.10 ($\tilde{\beta}=2.182$), and 0.16 ($\tilde{\beta}=2.148$), respectively. The values of various parameters are $R = 2500$ μm, $\Delta = 0.02$, $n_0^2 = 2.204$, $a = 25$ μm. [After Ref. 21]

$$\frac{d^2 z}{d\xi^2} = 2\left(\frac{dz}{d\xi}\right)\left(\frac{d\rho}{d\xi}\right)\left(1+\frac{\rho}{R}\right)^{-1}\frac{1}{R} + \frac{1}{2\tilde{\beta}^2}\left(1+\frac{\rho}{R}\right)^4 \frac{dn_2^2(z)}{dz} \tag{29}$$

which can be solved using Runge-Kutta's method for simultaneous second-order differential equations. The last term in Eq.(29) is actually absent in a bent slab waveguide; it appears for a bent fiber with a separable profile (see Sec. 5.4).

Meridional rays (for which $\tilde{l} = 0$) are confined to the plane of the bend and are shown in Figures 5.3 and 5.4. The rays are asymmetric about the axis of the waveguide, and as the radius of curvature increases this asymmetry decreases. For $R \to \infty$ the ray paths converge to a sinusoidal path of a straight waveguide.

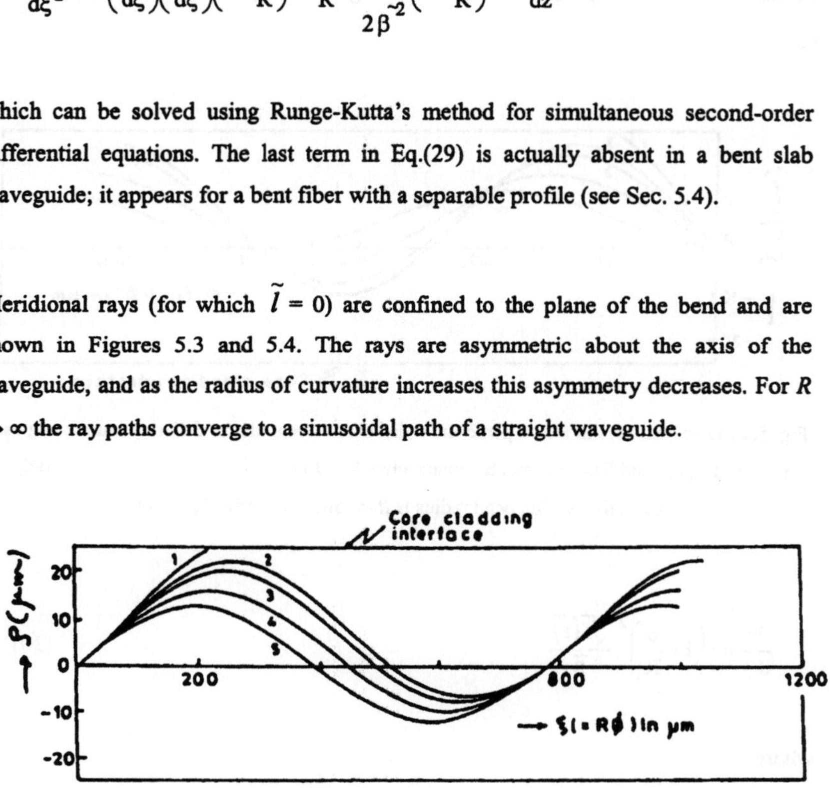

Fig. 5.3 Meridional rays in a bent parabolic-index slab waveguide for different radii of curvatures. Curves 1,2, 3, 4 and 5 correspond to R equal to 1500, 2000, 2500, 5000 and 25,000 μm, respectively. The launch angle is equal to 0.1 rad. Slab parameters are $n_0^2 = 2.204$, $\Delta = 0.02$, $a = 25$ μm. [After Ref. 21]

Ray paths in a bent fiber

We now consider a fiber whose refractive index profile is described by Eq.(12). We bend this fiber along the arc of a circle so that the y-axis of the straight fiber is parallel to the z-axis of the coordinate system described in Fig. 5.1. Now in this coordinate system, the fiber profile can be expressed as $n^2(r, z) = n_1^2(r) + n_2^2(z)$.

Using the analysis in Sec. 5.4, we have two invariants given by Eqs.(14) and (16). Consequently, the ray path can be described rigorously by Eqs.(17) and (20). Once again we have

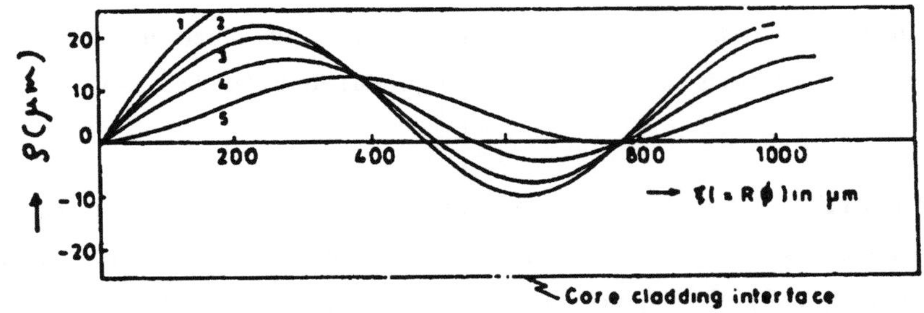

Fig. 5.4 Meridional rays in a bent parabolic-index slab waveguide for different launch angles. Curves 1,2, 3,4 and 5 correspond to launch angle θ = 0.16, 0.12, 0.10, 0.06, and 0.01 rad, respectively. The bend radius is $R = 2500\ \mu m$. [After Ref. 21]

$$\frac{dz}{d\xi} = \pm\left(1+\frac{\rho}{R}\right)^2 \frac{\sqrt{f(z)}}{\tilde{\beta}} \tag{30}$$

where

$$f(z) = n_2^2(z) + \tilde{l}^2 \tag{31}$$

and $d\rho/d\xi$ is given by Eq.(26). The exact ray paths can be obtained by solving the two simultaneous first-order differential equations given above. As before, it was found convenient to convert the above equations to two simultaneous second-order differential equations of the form given by Eqs.(28) and (29) where $dz/d\xi$ is now given by Eq.(30).

Figure 5.5 is a plot of the projection of the ray path on a cross section for skew rays obtained by launch conditions which correspond to a helical ray in a straight fiber. Figure 5.6 shows the corresponding variation in total distance from the waveguide axis. As the curvature increases, the rays become more and more skew. This can be explained as being due to the asymmetry induced by the bend.

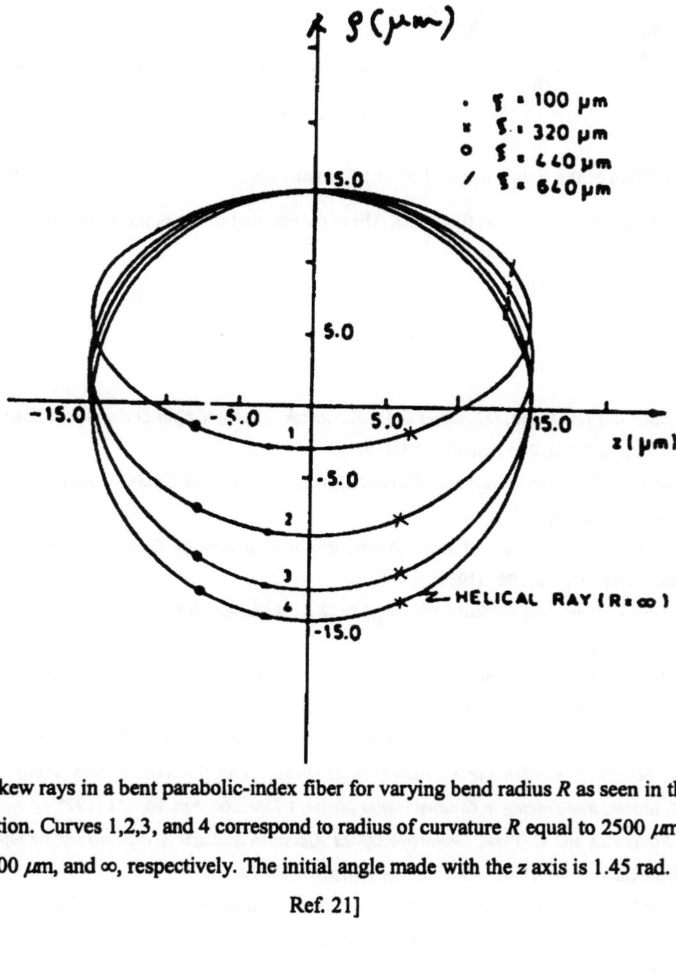

Fig. 5.5 Skew rays in a bent parabolic-index fiber for varying bend radius R as seen in the fiber cross section. Curves 1,2,3, and 4 correspond to radius of curvature R equal to 2500 μm, 5000 μm, 12500 μm, and ∞, respectively. The initial angle made with the z axis is 1.45 rad. [After Ref. 21]

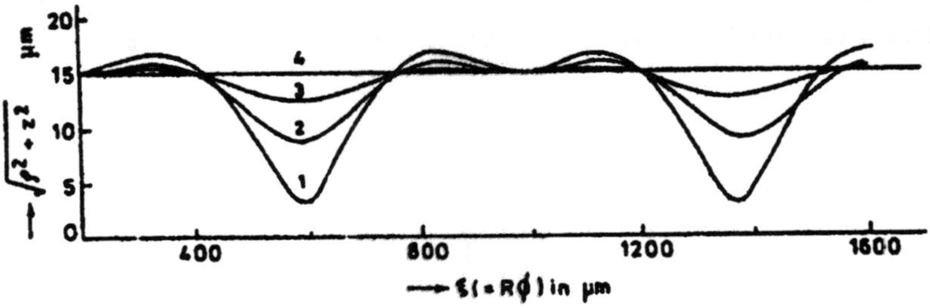

Fig. 5.6 Plots of the ray position $\sqrt{\rho^2 + z^2}$ inside the core as a function of distance along the parabolic-index bent fiber axis. These correspond to the plots shown in fig.5.5.

References

1. A.Gupta, K.Thyagarajan, I.C. Goyal and A.K. Ghatak, *Theoty of Fifth-Order Aberrations of Graded Index Media*, J. Opt. Soc. Am. **66**, 1320 (1976).
2. A. Rohra, K. Thyagarajan and A.K. Ghatak, *Aberrations in Curved Graded Index media*, J. Opt. Soc. Am. **69**, 300 (1979).
3. A. Sharma, D.V. Kumar and A.K. Ghatak, *Ray tracing through Graded-Index Media : a New Method*, Appl. Opt. **21**, 984 (1984).
4. A. Sharma, *Computing Optical Path Length in Gradient-Index Media : a Fast and Accurate Method*, Appl. Opt. **24**, 4367 (1985).
5. A. Sharma and A.K. Ghatak, *Ray Tracing in Gradient-Index Lenses : Computation of Ray-Surface Intersection*, Appl. Opt. **25**, 3409 (1986).
6. L. Montagnino, *Ray Tracing in Inhomogeneous Media*, J. Opt. Soc. Am. **58**, 1667 (1968).
7. W.H. Southwell, *Ray Tracing in Gradient-Index Media*, J. Opt. Soc. Am. **72**, 908 (1982).
8. D.T. Moore, *Ray Tracing in Gradient-Index Media*, J. Opt. Soc. Am. **65**, 451 (1975).
9. A. Ankiewicz and C. Pask, *Geometric Optics Approach to Light Acceptance and Propagation in Graded Index Fibers*, Opt. Quantum Electron, **9**, 87 (1977).
10. D. Bertilone and C..Pask, *Exact Ray Paths in a Graded-Index Taper*, Appl. Opt. **26**, 1189 (1987).
11. E.W. Marchand, *Gradient Index Optics* , Academic Press, New York, (1978).
12. M.J. Nadeau, *Image Analysis of Curved Gradient-Index Rods*, M.S. Thesis, Institute of Optics, U. Rochester (1984).
13. A.W. Snyder and J.D. Love, *Optical Waveguide Theory* ,Chapman & Hall, London, (1983).
14. C. Winkler, J.D. Love and A.K. Ghatak, *Loss Calculations in Bent Multimode Optical Waveguides*, Opt. Quantum Electron. **11**, 173 (1979).
15. R.K. Luneberg, *Mathematical Theory of Optics*, Univ. of California, Berkeley, (1964).
16. A.K. Ghatak and K. Thyagarajan, *Contemporary Optics*, Plenum, New York, (1978).

17. K. Thyagarajan, M.R. Shenoy and A.K. Ghatak, *Accurate Numerical Method for the Calculation of Bending Loss in Optical Waveguides using a Matrix Approach*, Opt. Lett. **12**, 296 (1987).
18. A. Kumar, K. Thyagarajan and A.K. Ghatak, *Analysis of Rectangular Core Dielectric Waveguides: An Accurate Perturbation Approach*, Opt. Lett. **8**, 63 (1983).
19. A. Kumar, R.K. Varshney and K. Thyagarajan, *Birefringence Calculations in Elliptical Core Optical Fibers*, Electron. Lett. **20**, 112 (1984).
20. J.B. Scarborough, *Numerical Mathematical Analysis*, Oxford and IBH Publishing, Co., London, (1966).
21. A. Ghatak, E. Sharma, and J. Kompella, *Exact ray paths in bent waveguides*, Applied Optics, **27**, 3180, (1988).

6. THE OPTICAL HAMILTONIAN AND STUDY OF PARAXIAL LENS OPTICS

The optical Hamiltonian

Analogous to the case in classical mechanics, one can also develop the Hamiltonian formulation in optics. This approach will be used in Chapter 7 to calculate explicit expressions for various aberration coefficients. In the Hamiltonian formulation, we have first to define the generalized momenta p and q by the relation[1]

$$p = \frac{\partial L}{\partial \dot{x}}$$
$$q = \frac{\partial L}{\partial \dot{y}} \qquad (1)$$

where, as before, dots represent differentiation with respect to z and L is the optical Lagrangian. On substituting the value of L from Eq.(8) of Chapter 3, we find

[1] In classical mechanics q's represent the generalized coordinates and p's the corresponding generalized momenta. Notice that here both p and q are canonical momenta corresponding to x and y, respectively.

$$p = \frac{n\dot{x}}{\sqrt{1+\dot{x}^2+\dot{y}^2}} = n\frac{dx}{ds}$$
$$q = \frac{n\dot{y}}{\sqrt{1+\dot{x}^2+\dot{y}^2}} = n\frac{dy}{ds} \tag{2}$$

Since *dx/ds* and *dy/ds* represent the direction cosines of the ray at the point (*x, y, z*) along the *x* and *y* directions, *p*(= *n dx/ds*) and *q*(= *n dy/ds*) are termed the optical direction cosines of the ray. We now define the optical Hamiltonian *H*, in terms of *L*, through the relation

$$H = p\dot{x} + q\dot{y} - L(x, y, \dot{x}, \dot{y}, z) \tag{3}$$

Thus

$$dH = \left(p - \frac{\partial L}{\partial \dot{x}}\right)d\dot{x} + \left(q - \frac{\partial L}{\partial \dot{y}}\right)d\dot{y} + \dot{x}dp + \dot{y}dq$$
$$- \frac{\partial L}{\partial x}dx - \frac{\partial L}{\partial y}dy - \frac{\partial L}{\partial z}dz \tag{4}$$
$$= \dot{x}dp + \dot{y}dq - \dot{p}dx - \dot{q}dy - \frac{\partial L}{\partial z}dz$$

From Eq.(4) it is clear that *H* is a function of *x, y, p, q*, and *z* and the following equations, Hamilton's equations, also follow readily:

$$\dot{x} = \frac{\partial H}{\partial p} \quad \text{and} \quad \dot{y} = \frac{\partial H}{\partial q} \tag{5}$$

$$\dot{p} = -\frac{\partial H}{\partial x} \quad \text{and} \quad \dot{q} = -\frac{\partial H}{\partial y} \tag{6}$$

$$\frac{\partial H}{\partial z} = -\frac{\partial L}{\partial z} \tag{7}$$

These equations from the basic equations of the Hamiltonian formulation. Given a Hamiltonian *H*, i.e., given a refractive index function *n(x,y,z)*, the above equations allow us to calculate the ray path. If we substitute the expressions for *L*, *p* and *q* from Eq.(8) of Chapter 3 and Eq.(2) into Eq.(3), we obtain

$$H = \frac{n\dot{x}}{\sqrt{1+\dot{x}^2+\dot{y}^2}}\dot{x} + \frac{n\dot{y}}{\sqrt{1+\dot{x}^2+\dot{y}^2}}\dot{y} - n\sqrt{1+\dot{x}^2+\dot{y}^2}$$

$$= -\frac{n}{\sqrt{1+\dot{x}^2+\dot{y}^2}}$$

We also note that

$$\sqrt{n^2-p^2-q^2} = \sqrt{n^2 - \frac{n^2\dot{x}^2}{1+\dot{x}^2+\dot{y}^2} - \frac{n^2\dot{y}^2}{1+\dot{x}^2+\dot{y}^2}}$$

$$= \frac{n}{\sqrt{1+\dot{x}^2+\dot{y}^2}}$$

Thus

$$H = -\sqrt{n^2(x,y,z)-p^2-q^2} \qquad (8)$$

Application of the Hamiltonian formulation to the study of paraxial lens optics

Using Hamilton's equations, we will trace the rays in a rotationally symmetric optical system. By rotational symmetry, we imply that the properties of the system are the same on the circumference of any circle whose center lies on an axis which is known as the symmetry axis of the system. A simple example is a coaxial system of lenses. Even for a rotationally symmetric system, since the Hamiltonian *H* is an irrational function of *x*, *y*, *p*, *q*, and *z*, it is difficult to solve Hamilton's equations and one has to look for approximate solutions. The lowest-order approximation would lead us to paraxial optics or Gaussian optics, which is concerned with rays that travel close to the axis of the system and make small angles with it. Under this approximation it will be shown that perfect images can be formed. Deviations from these determine the aberrations of the system. In this section we will show how the Hamiltonian formulation can be used to yield simple results for refracting surfaces and lenses, which are normally obtained by application of Snell's law. We will consider a few representative examples to show the applicability of this method. In Chapter 7 we will use the same formulation to calculate explicit expressions for the aberrations introduced by rotationally symmetric systems.

Since for a rotationally symmetric system the refractive index depends on the value of (x^2+y^2) (rather than on x and y independently) we may write the Hamiltonian in the form

$$H = -\sqrt{n^2(u,z) - v} \tag{9}$$

where

$$\begin{aligned} u &= x^2 + y^2 \\ v &= p^2 + q^2 \end{aligned} \tag{10}$$

In the paraxial approximation, u and v are small quantities and hence we may make a Taylor series expansion of the Hamiltonian in ascending powers of u and v, retaining only first-order terms:

$$H(u,v,z) = H_0(z) + [H_1(z)u + H_2(z)v] + \cdots \tag{11}$$

where

$$H_1(z) = \left.\frac{\partial H}{\partial u}\right|_{u=0,v=0}$$
$$H_2(z) = \left.\frac{\partial H}{\partial v}\right|_{u=0,v=0} \tag{12}$$

In order to calculate the aberrations, higher-order terms in Eq. (11) would have to be retained (see chapter 7).

Since our system possesses rotational symmetry, we need consider only the set of equations in x and p; the equations for y and q would follow from analogy. From Eq.(10), we obtain

$$\frac{\partial}{\partial x} = 2x\frac{\partial}{\partial u}$$
$$\frac{\partial}{\partial p} = 2p\frac{\partial}{\partial v} \tag{13}$$

Thus, Hamilton's equations [Eqs.(5) and (6)] become

$$\dot{x} = \frac{dx}{dz} = 2p\frac{\partial H}{\partial v} = 2H_2 p$$
$$\dot{p} = \frac{dp}{dz} = -2x\frac{\partial H}{\partial u} = -2H_1 x \tag{14}$$

where we have used Eq.(11). Here x and p correspond to the paraxial approximation. Using Eqs.(8) and (12), we obtain

$$H_1 = \left.\frac{\partial n}{\partial u}\right|_{u=0,v=0}$$
$$H_2 = \frac{1}{2n(0,z)} \tag{15}$$

where $n(0, z)$ is the refractive index along the axis. We will now make use of the above formulation to study the imaging properties of some simple optical systems, like a single refracting surface, a thin lens, and a thick lens. Other complicated optical systems can be analyzed by using these formulas in conjunction.

A single refracting surface

Let us first consider a spherical refracing surface (of radius of curvature R) separating two homogeneous media of refractive indices n_1 and n_2 (see fig. 6.1).

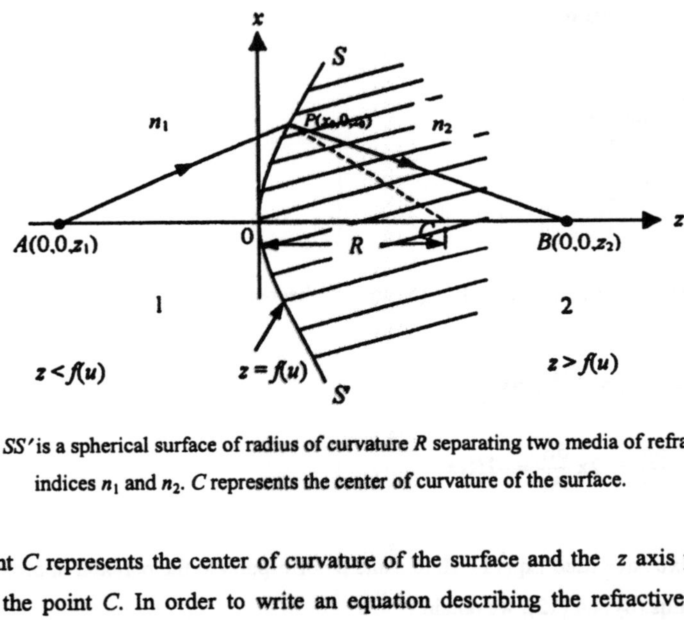

Fig. 6.1 SS' is a spherical surface of radius of curvature R separating two media of refractive indices n_1 and n_2. C represents the center of curvature of the surface.

The point C represents the center of curvature of the surface and the z axis passes through the point C. In order to write an equation describing the refractive-index variation, we must first find the equation of the spherical surface. Let (x,y,z) be the coordinates of an arbitrary point on the surface; the origin is assumed to be at the point O. Since the refracting surface is a portion of a sphere, we must have

$$x^2 + y^2 = z(2R - z) \tag{16}$$

Thus

$$z = R\sqrt{1 \pm \left(1 - \frac{u^2}{R^2}\right)} \cong \frac{u}{2R} + \frac{u^2}{8R^3} + \cdots = f(u) \tag{17}$$

where we have chosen the solution that makes z go to zero as $u \rightarrow 0$. (Positive values of R correspond to a convex surface and negative values of R correspond to a concave surface.) Clearly, a point whose z coordinate satisfies the inequality $z > f(u)$ lies to the right of the surface. Similarly if $z < f(u)$ then the point will lie to the left of the surface. Hence the refractive index variation can be written in the form

$$n(x, y, z) = n_1 \Theta(f(u) - z) + n_2 \Theta(z - f(u)) \tag{18}$$

where $\Theta(x)$ is the unit step function, defined by the equation

$$\Theta(x) = \begin{cases} 1 & x > 0 \\ 0 & x < 0 \end{cases} \tag{19}$$

Thus

$$H_1 = -\left.\frac{\partial n}{\partial u}\right|_{u=0} = \frac{n_2 - n_1}{2R}\delta(z) \tag{20}$$

where we have used

$$\frac{d\Theta(x)}{dx} = \delta(x) \tag{21}$$

and $\delta(x)$ is the Dirac delta function. Thus

$$\frac{dp}{dz} = -2H_1 x = -(n_2 - n_1)\frac{\delta(z)}{R}x \tag{22}$$

It can immediately be seen that for $z < 0$ or for $z > 0$, $dp/dz = 0$, which shows that the rays connecting any two points lying in the same medium are straight lines. However, when the ray hits the refracting surface, it undergoes an abrupt change in its slope as can be seen by integrating Eq.(22) from $z = -\varepsilon$ to $z = +\varepsilon$:

$$\int_{-\varepsilon}^{\varepsilon} \frac{dp}{dz} dz = -\int_{-\varepsilon}^{\varepsilon} \frac{n_2 - n_1}{R} \delta(z) x dz \tag{22}$$

or

$$p_2 - p_1 = -\frac{n_2 - n_1}{R} x_0 \tag{23}$$

where x_0 is the value of the x coordinate of the ray calculated at the refracting surface and p_1 and p_2 are the optical direction cosines in media 1 and 2, respectively. Further,

$$p = \frac{1}{2H_2} \dot{x} = n(0,z) \frac{dx}{dz} = \begin{cases} n_1 \dfrac{dx}{dz} & \text{in medium 1} \\ n_2 \dfrac{dx}{dz} & \text{in medium 2} \end{cases} \tag{24}$$

Let us consider a ray that starts from the point $A(0,0,z_1)$ and gets refracted at the point $P(x_0,0,z_0)$ as shown in Fig. 6.1; obviously z_1 is a negative quantity. (We are using the analytical geometry convention). Thus, using Eq.(24) we get

$$\begin{aligned} p_1 &= n_1 \frac{x_0}{z_0 - z_1} \cong -n_1 \frac{x_0}{z_1} \\ p_2 &\cong -n_2 \frac{x_0}{z_2} \end{aligned} \tag{25}$$

Substituting the above values in Eq.(23), we obtain

$$\frac{n_2}{z_2} - \frac{n_1}{z_1} = \frac{n_2 - n_1}{R} \tag{26}$$

which is the required formula for a single refracting surface. It can be seen from the above equation that in the realm of paraxial optics, for a given object point, the position of the image point is dependent only on n_1, n_2, and R. Thus the image formed under paraxial optics is an ideal image. The primary and the secondary focal lengths are $-n_1R/(n_2 - n_1)$ and $n_2R/(n_2 - n_1)$.

Thin lens

A lens is called thin if a ray striking the first surface emerges at approximately the same height from the other surface. Consider a thin lens of refractive index n_1 placed between two media of refractive indices n_0 and n_2 as shown in Fig.6.2a.

Let the radii of curvatures of the two surfaces forming the lens be R_1 and R_2, respectively. If p_0 is the direction cosine of the incident ray, then

$$\begin{aligned} p_1 - p_0 &= -\frac{n_1 - n_0}{R_1} x_0, \\ p_2 - p_1 &= \frac{n_1 - n_2}{R_2} x_0 \end{aligned} \tag{27}$$

where p_1 and p_2 are the optical direction cosines of the ray in media 1 and 2 and x_0 is the height at which the ray strikes the lens. Thus

$$p_2 - p_0 = \left(\frac{n_1 - n_2}{R_2} - \frac{n_1 - n_0}{R_1} \right) x_0 \tag{28}$$

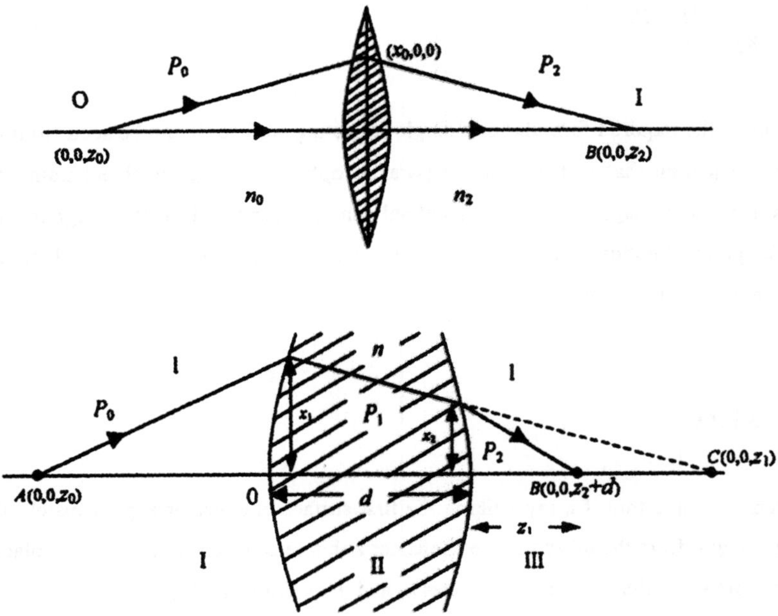

Fig. 6.2 (a) A thin lens made of a material of refractive index n_1 placed between media of refractive indices n_0 and n_2. (b) A thick lens made of a material of refractive index n placed in air.

Since

$$p_0 = -n_0 \frac{x_0}{z_0}$$
$$p_2 = -n_2 \frac{x_0}{z_2} \tag{29}$$

(see Eq. (25)), we obtain

$$\frac{n_0}{z_0} - \frac{n_2}{z_2} = -\left(\frac{n_1 - n_0}{R_1} - \frac{n_1 - n_2}{R_2}\right) \tag{30}$$

which is the thin-lens formula. In most cases, the two media on both sides of the lens are the same, i.e., $n_2 = n_0$, and if $n = n_1/n_0$ we get from Eq.(30)

$$\frac{1}{z_2} - \frac{1}{z_0} = (n-1)\left(\frac{1}{R_1} - \frac{1}{R_2}\right) = \frac{1}{f} \tag{31}$$

where f represents the focal length of the lens.

Thick lens

If the thickness of the lens is not negligible compared to other parameters then one has a thick lens. Such a lens is shown in Fig.6.2(b). Let R_1 and R_2 be the radii of curvature of the two surfaces and let d be the thickness of the lens on the axis. For simplicity let us consider the media surrounding the lens to be of refractive index unity and let n represent the refractive index of the material of the lens. Let the vertex of the first surface represent the origin of the coordinate system. Consider a ray that starts from point A with coordinates $(0,0,z_0)$ and intersects the first surface at a height x_1 from the axis. After refraction let the ray intersect the second surface at a distance x_2 from the axis. The emerging ray passes through the point $B(0,0,z_2+d)$, where z_2 and d have been defined in Fig. 6.2(b). If p_0, p_1, and p_2 represent the values of p in the three regions I, II and III, then corresponding to Eq. (27) we have at the first and second surfaces

$$\begin{aligned} p_1 - p_0 &= -\frac{n-1}{R_1}x_1 \\ p_2 - p_1 &= \frac{n-1}{R_2}x_2 \end{aligned} \tag{32}$$

(For a thin lens, we would have $x_1 = x_2$.) From the definition of optical direction cosine, we can write

$$x_2 - x_1 = \frac{p_1}{n} d \tag{33}$$

Using Eqs. (32) and (33) we obtain

$$p_0\left(1+\frac{n-1}{n}\frac{d}{R_2}\right) - p_2 = \frac{n-1}{R_1} x_1\left(1+\frac{n-1}{n}\frac{d}{R_2}\right) - \frac{n-1}{R_2} x_1 \tag{34}$$

Let us consider the case when $p_2 = 0$, i.e., the ray emerging from the lens is parallel to the axis. This will correspond to a particular object position shown as F_1 in Fig. 6.3(a).

A plane normal to the axis and passing through the point of intersection of the incident and the emergent rays is known as the primary principal plane and is shown as P_1 in the figure. The point F_1 is known as the primary focal point and its distance from the principal plane P_1 (which is denoted by f_1) is known as the primary focal length. It can easily be seen from the figure that

$$p_0 = \frac{x_2}{F_1 N} = \frac{x_2}{(-f_1)} \tag{35}$$

and using the fact that $p_2 = 0$, we obtain $f_1 = -\ nR_1R_2/D$, where

$$D = (n-1)[n(R_2 - R_1) + (n-1)d] \tag{36}$$

Similarly, the second focal length f_2 (see Fig.6.3(b)) is nR_1R_2/D. The magnitudes of the two focal lengths happen to be equal because of the medium being the same on both sides of the lens. The two planes P_1 and P_2 shown in the figure are called the principal planes of the lens. They are also called unit magnification planes.

The distances of the principal planes from the respective vertices can be easily determined. From Fig. 6.3(a), we find by simple application of geometry

$$t_1 = f_1\left(1 - \frac{x_1}{x_2}\right) = -(n-1)\frac{dR_1}{D} \tag{37}$$

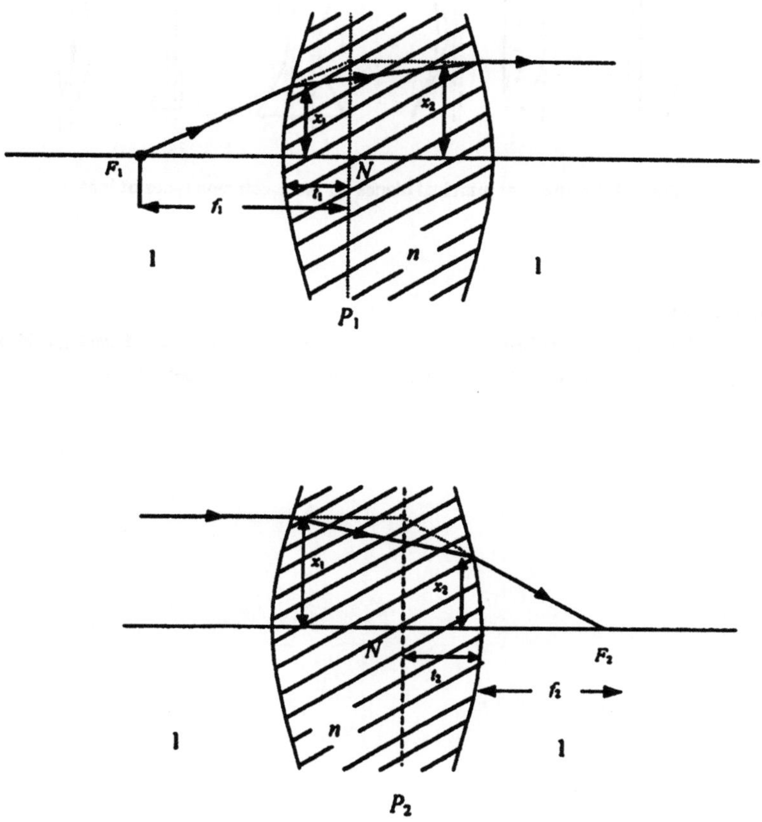

Fig.6.3 (a) P_1 and P_2 represent the two principal planes of the lens. The focal lengths are measured from the respective principal planes. The index of refraction of the lens is n and that of the surrounding medium is 1.

Similarly, we can obtain (see Fig. 6.3(b))

$$t_2 = (n-1)\frac{dR_2}{D} \tag{38}$$

The positions of the principal planes for some common types of lenses are shown in Fig. 6.4.

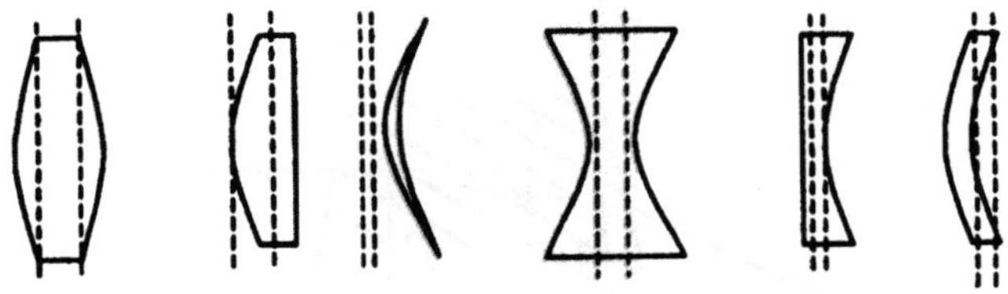

Fig. 6.4 Positions of principal planes for some common types of lenses.

References

1. R.K. Luneberg, *Mathematical Theory of Optics*, Univ. California Press, Berkeley, (1964).
2. M. Born and E. Wolf, *Principles of Optics*, Pergamon Press, Oxford, (1975).

7. GEOMETRICAL THEORY OF THIRD-ORDER ABERRATIONS

Introduction

In Chapter 6 we used the Hamiltonian formulation to trace rays in some rotationally symmetric optical systems. They were derived under the paraxial approximation, i.e., the rays forming the image were assumed to lie infinitesimally close to the axis and to make infinitesimally small angles with it. It was found that the images of point objects were perfect, i.e., all rays starting from a given object point were found to intersect at *one* point, which is the image point. Such an image is called an ideal image. In general, rays that make large angles with the *z* axis or travel at large distances from the *z* axis do not intersect at one point. This phenomenon is known as aberration. In this chapter we will use the Hamiltonian formulation as developed in Ref. 1 to derive explicit expressions for aberrations in rotationally symmetric systems. Detailed general discussion of aberrations is given elsewhere [2,3].

In Section 'Expression of third order aberration' we derive explicit expressions for third-order aberrations in terms of two specific paraxial rays. In Section 'Physical significance of coefficients A,B,C,D &E' we discuss how the different aberration coefficients obtained in Section 'Expression of third order aberration' describe the five Seidel aberrations, namely, spherical aberration, coma, astigmatism, curvature of field, and distortion. After obtaining the explicit dependence of H_i and H_{ij} in terms of the refractive index function, in Section 'The coefficient H_{ij} in term of refractive index variation' we obtain explicit expressions for the aberrations of a rotationally symmetric graded index medium, whose paraxial property we have already studied in Chapter 6. In Section 'Aberrations in systems possessing finite discontinuities in refractive index' we obtain explicit expressions for the aberration coefficients in optical systems possessing finite discontinuities in refractive index, the refractive index between two discontinuities being constant. Examples of such optical systems are systems made up of lenses. Finally, in Section on 'Chromatic aberration' we discuss the chromatic aberrations of optical systems.

Expressions for third-order aberration

For a rotationally symmetric system, the Hamiltonian is given by

$$H = -\sqrt{n^2(U, z) - V} \tag{1}$$

where $U = X^2 + Y^2$, $V = P^2 + Q^2$, (X, Y, z) represents the position coordinate of any point on the ray, and (P, Q) are the optical direction cosines of the ray[1]. Hamilton's equations are

$$\begin{aligned} \frac{dX}{dz} &= 2P\frac{\partial H}{\partial V} \\ \frac{dP}{dz} &= -2X\frac{\partial H}{\partial U} \end{aligned} \tag{2}$$

To specify a ray one has to specify either the values of X, Y, P, and Q in any plane or the values of X and Y in two planes. We will choose the second type of boundary conditions (similar results can also be obtained by using the first set of boundary conditions). Let $z = z_0$ represent the object plane and $z = \zeta$ any other judiciously chosen reference plane; we will choose the plane $z = \zeta$ to contain the exit pupil. In

[1] The expression for the Hamiltonian is derived in Chapter 6. Here we are using capital letters for the transverse coordinates of a general ray to differentiate it from paraxial rays.

general, a ray is completely specified if we know $X(z_0)$, $Y(z_0)$, $X(\zeta)$ and $Y(\zeta)$. For a particular ray, let these be given by

$$\begin{aligned} X(z_0) &= x_0 \\ X(\zeta) &= \xi \end{aligned} \tag{3}$$

$$\begin{aligned} Y(z_0) &= y_0 \\ Y(\zeta) &= \eta \end{aligned} \tag{4}$$

If we solve Eq. (2) under these boundary conditions, then X, Y, P and Q will be general functions of x_0, y_0, ξ and η. Paraxial rays would correspond to infinitesimal values of x_0, y_0, ξ and η. For any general ray, we expand X, Y, P and Q in ascending powers of x_0, y_0, ξ and η to get

$$\begin{aligned} X &= X_1 + X_2 + X_3 + \cdots \\ P &= P_1 + P_2 + P_3 + \cdots \end{aligned} \tag{5}$$

and similar equations for Y and Q. The subscripts represent the degree of the polynomial; for example, X_1 is linear in x_0, y_0, ξ and η; X_2 is quadratic in x_0, y_0, ξ and η. Note that there is no term like X_0 or P_0; this is because of the fact that a ray travelling along the symmetry axis (i.e., $X = Y = 0$) of a rotationally symmetric system travels undeviated. Hence if x_0, y_0, ξ and η are zero, then (X, Y, P, Q) are zero everywhere. For infinitesimal values of x_0, y_0, ξ and η i.e., for a paraxial ray, X_2, X_3, etc., can be neglected in Eq. (5). Hence X_1 represents the paraxial value of X. Similarly P_1, Y_1 and Q_1 represent paraxial values of P, Y and Q respectively.

We will now show how the rotational symmetry of the system prohibits the even-order terms in the expansions given by Eq. (5). Since the system is rotationally symmetric, if a ray specified by the coordinates (x_0, y_0, ξ and η) has values $X(z_1)$, $Y(z_1)$, $P(z_1)$ and $Q(z_1)$ in some plane $z = z_1$, then another ray specified by ($-x_0$, $-y_0$, $-\xi$ and $-\eta$) should have values $-X(z_1)$, $-Y(z_1)$, $-P(z_1)$ and $-Q(z_1)$ in the same plane $z = z_1$ (see Fig. 7.1). If we try to impose this property in Eq. (5), we find that the quantities $X_1, X_3, \ldots$ change sign, while the quantities $X_2, X_4, \ldots$ do not. Hence it follows that the even-order terms should be absent from the expansions; thus Eq. (5) becomes

$$\begin{aligned} X &= X_1 + X_3 + \cdots \\ P &= P_1 + P_3 + \cdots \end{aligned} \tag{6}$$

X_3 represents the third-order correction to X_1 and is called the third-order aberration. It is an aberration term because if it were absent (and $X_5, X_7, \ldots$ etc., were also absent) then as will be shown shortly, the images of every point object would be an exact point, and hence would represent an ideal image. Similarly, X_5 is called the fifth-order aberration. X_7 the seventh-order aberration, etc.

In order to calculate the aberrations, we make a Taylor expansion of H given by Eq. (1) in ascending powers of U and V about $U=0$, $V=0$, to get

$$H = H_0 + H_1U + H_2V + \frac{1}{2}(H_{11}U^2 + 2H_{12}UV + H_{22}V^2) + \cdots \tag{7}$$

where H_i and H_{ij} are the Taylor coefficients and are all functions of z. Equation (7) may be compared with Eq. (11) of chapter 6 where terms up to first order in U and V were retained to obtain paraxial optics.

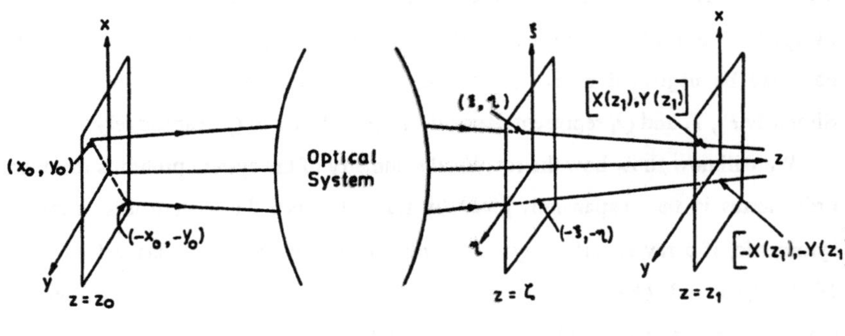

Figure 7.1 A rotationally symmetric optical system.

Substituting the expansions of X, P, and H from Eq. (6) and (7) into Eq. (2), we get

$$(\dot{X}_1+\dot{X}_3+\cdots)=2(P_1+P_3+\cdots)(H_2+H_{12}U+H_{22}V+\cdots) \tag{8}$$

where as before dots represent derivatives with respect to z. Now,

$$\begin{aligned}U=X^2+Y^2&=(X_1+X_3+\cdots)^2+(Y_1+Y_3+\cdots)^2\\&=U_1+2(X_1X_3+Y_1Y_3)+\text{higher order terms}\end{aligned} \tag{9}$$

where $U_1=X_1^2+Y_1^2$. Similarly,

$$V=V_1+2(P_1P_3+Q_1Q_3)+\text{higher order terms} \tag{10}$$

where $V_1=P_1^2+Q_1^2$. Thus

$$\begin{aligned}(\dot{X}_1+\dot{X}_3+\cdots)=2(P_1+P_3+\cdots)\times\\ \begin{Bmatrix}H_2+H_{12}[U_1+2(X_1X_3+Y_1Y_3)+\cdots]+\\ H_{22}+[V_1+2(P_1P_3+Q_1Q_3)+\cdots]\end{Bmatrix}\end{aligned} \tag{11}$$

Equating terms of equal order we obtain

$$\dot{X}_1=2H_2P_1 \tag{12}$$

$$\dot{X}_3=2H_2P_3+2(H_{12}U_1+H_{22}V_1)P_1 \tag{13}$$

Similarly,

$$\dot{P}_1=-2H_1X_1 \tag{14}$$

$$\dot{P}_3=-2H_1X_3-2(H_{11}U_1+H_{12}V_1)X_1 \tag{15}$$

Equations (12) and (14) represent paraxial equations. Equations (13) and (15) are equations that determine the third-order aberrations in the system. The corresponding equations for Y_1, Q_1, Y_3 and Q_3 would be

$$
\begin{aligned}
\dot{Y}_1 &= 2H_2Q_1 \\
\dot{Y}_3 &= 2H_2Q_3 + 2(H_{12}U_1 + H_{22}V_1)Q_1 \\
\dot{Q}_1 &= -2H_1Y_1 \\
\dot{Q}_3 &= -2H_1Y_3 - 2(H_{11}U_1 + H_{12}V_1)Y_1
\end{aligned}
\tag{16}
$$

Given an optical system, we know the refractive index n as a function of U and z. Knowing n, the coefficients H_i and H_{ij} can be calculated. Then the paraxial equations (12) and (14) can be solved. This solution can then be substituted in Eqs. (13) and (15) to get the expression for X_3. An analogous procedure can be adopted for Y_3.

We will now use two specific paraxial rays to get explicit expressions for X_3 and Y_3 directly in terms of the system parameters. The first paraxial ray, called the axial ray, satisfies the boundary conditions

$$
\begin{aligned}
X_1(z_0) &= 0 \\
Y_1(z_0) &= 0 \\
X_1(\zeta) &= 1 \\
Y_1(\zeta) &= 0
\end{aligned}
\tag{17}
$$

The solution of the paraxial equations under this boundary condition would be represented by $X_1(z) = g(z)$ and $P_1(z) = \vartheta(z)$ such that $g(z_0) = 0$, $g(\zeta) = 1$ (see Fig. 7.2). The second paraxial ray, called the field ray, satisfies the boundary conditions

$$
\begin{aligned}
X_1(z_0) &= 1 \\
Y_1(z_0) &= 0 \\
X_1(\zeta) &= 0 \\
Y_1(\zeta) &= 0
\end{aligned}
\tag{18}
$$

The solution of the paraxial equations under this set of boundary conditions would be represented by $X_1(z) = G(z)$ and $P_1(z) = \theta(z)$ such that $G(z_0) = 1$, $G(\zeta) = 0$ (see Fig. 7.2).

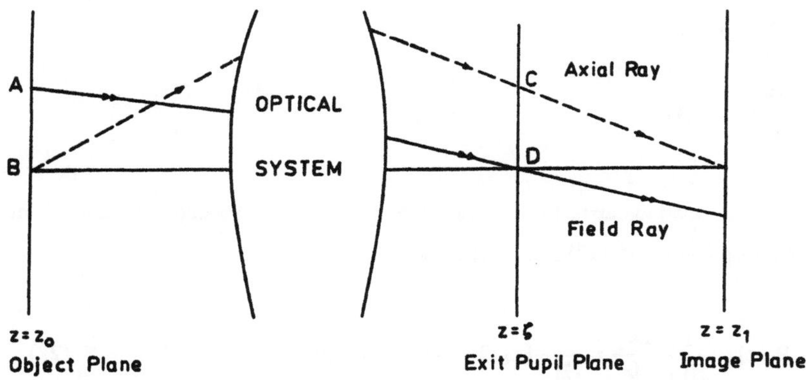

Figure 7.2 The dashed and solid curves correspond to the axial and field rays respectively. *AB* and *CD* are of unit length.

Thus $g(z)$, $G(z)$, $\vartheta(z)$ and $\theta(z)$ form two linearly independent solutions of the set of paraxial equations. Hence solutions of eqs. (12) and (14) satisfying any general boundary condition can be expressed as linear combinations of these. This is so because we have a set of coupled first-order differential equations, which is equivalent to a second-order differential equation. Thus for a ray that satisfies the boundary conditions specified by eqs. (3) and (4), the paraxial solution is given by

$$\begin{aligned} X_1(z) &= x_0 G(z) + \xi g(z) \\ Y_1(z) &= y_0 G(z) + \eta g(z) \end{aligned} \tag{19}$$

This can be verified by noting that $G(z_0) = 1$ and $g(z_0) = 0$, and $G(\zeta) = 0$, $g(\zeta) = 1$. We also have

$$\frac{dg}{dz} = 2H_2\vartheta$$
$$\frac{d\vartheta}{dz} = -2H_1 g \tag{20}$$

$$\frac{dG}{dz} = 2H_2\theta$$
$$\frac{d\theta}{dz} = -2H_1 G \tag{21}$$

The above equations arise because (g, ϑ) and (G, θ) are special solutions of the paraxial equation (eqs. (12) and (14)). Now,

$$P_1(z) = \frac{1}{2H_2}\frac{dX_1}{dz} = \frac{1}{2H_2}(x_0\dot{G} + \xi\dot{g}) = x_0\theta + \xi\vartheta \tag{22}$$

Similarly,

$$Q_1(z) = y_0\theta + \eta\vartheta \tag{23}$$

We are interested in calculating the aberration (i.e., the value of X_3) in the paraxial image plane defined in the following manner: Let us consider an axial point in the object plane $z = z_0$. Then a ray leaving this point and travelling along the axis of symmetry of the system would not be deviated. Any other paraxial ray starting from this point would be represented by a multiple of $g(z)$. The image would be formed at the intersection of these two rays, i.e., at a value of z, say $z = z_1$, such that $g(z_1) = 0$, $z_1 \neq z_0$. This would then represent the paraxial image plane (see fig. 7.2).

Since $g(z_1) = 0$, therefore in the image plane the paraxial solution (see Eq. (19)) would be simply

$$X_1(z_1) = x_0 G(z_1) \tag{24}$$

Since $G(z_1)$ is independent of x_0, y_0, ξ and η, Eq. (24) says that the paraxial image of a point (which is at a distance x_0 from the z axis) in the object plane would be formed at the point that is at a distance $X_1(z_1)$ from the z axis in the image plane $z = z_1$, i.e., all rays emanating from the point $(x_0, 0)$ in the plane $z = z_0$ would meet at $[X_1(z_1), 0]$, irrespective of the values of ξ and η. Thus the image would be an ideal image.

Let us now calculate the aberrations X_3 and Y_3 of a ray satisfying the boundary conditions given by eqs. (3) and (4) in terms of the functions $g(z)$, $\vartheta(z)$, $G(z)$ and $\Theta(z)$. If we solve the paraxial equations with the boundary conditions

$$\begin{aligned} X_1(z_0) &= x_0 \\ X_1(\zeta) &= \xi \end{aligned} \tag{25}$$

$$\begin{aligned} Y_1(z_0) &= y_0 \\ Y_1(\zeta) &= \eta \end{aligned} \tag{26}$$

then clearly we have

$$\begin{aligned} X_3(z_0) &= X_5(z_0) = \cdots = 0 \\ Y_3(z_0) &= Y_5(z_0) \cdots = 0 \end{aligned} \tag{27}$$

The solution of the paraxial equations under the boundary conditions given by eqs. (25) and (26) is indeed given by eqs. (19), (22) and (23). Substituting for $2H_2$ from Eq. (20) into Eq. (13) we get

$$\dot{X}_3\vartheta - \dot{g}P_3 = 2(H_{12}U_1 + H_{22}V_1)P_1\vartheta \tag{28}$$

Similarly, substituting for $2H_1$ from Eq. (20) in Eq. (15) we get

$$X_3\dot{\vartheta} - g\dot{P}_3 = 2(H_{11}U_1 + H_{12}V_1)X_1 g \tag{29}$$

Adding eqs. (28) and (29) we get

$$\frac{d}{dz}(X_3\vartheta - gP_3) = 2[(H_{12}U_1 + H_{22}V_1)P_1\vartheta + (H_{11}U_1 + H_{12}V_1)X_1 g] \tag{30}$$

Integrating the above equation from the object plane $z = z_0$ to the image plane $z = z_1$, we get

$$(X_3\vartheta - gP_3)\Big|_{z_0}^{z_1} = 2\int_{z_0}^{z_1}[(H_{12}U_1 + H_{22}V_1)P_1\vartheta + (H_{11}U_1 + H_{12}V_1)X_1 g]dz \tag{31}$$

Using the fact that $g(z_0) = 0, X_3(z_0) = 0, \; g(z_1) = 0$, we get

$$X_3(z_1) = \frac{2}{\vartheta(z_1)}\int_{z_0}^{z_1}[(H_{12}U_1 + H_{22}V_1)P_1\vartheta + (H_{11}U_1 + H_{12}V_1)X_1 g]dz \tag{32}$$

which represents the aberration of the ray along the x direction in the paraxial image plane. Similarly,

$$Y_3(z_1) = \frac{2}{\vartheta(z_1)}\int_{z_0}^{z_1}[(H_{12}U_1 + H_{22}V_1)Q_1\vartheta + (H_{11}U_1 + H_{12}V_1)Y_1 g]dz \tag{33}$$

which represents the aberration of the ray along the y direction in the paraxial image plane. We write X_3 and Y_3 explicitly in terms of g, G, ϑ, and θ by observing that

$$\begin{aligned} U_1 &= X_1^2 + Y_1^2 = [x_0 G(z) + \xi g(z)]^2 + [y_0 G(z) + \eta g(z)]^2 \\ &= rG^2 + 2tGg + sg^2 \end{aligned} \tag{34}$$

where

$$r = x_0^2 + y_0^2$$
$$t = x_0\xi + y_0\eta$$
$$s = \xi^2 + \eta^2$$

Similarly,

$$\begin{aligned} V_1 = P_1^2 + Q_1^2 &= [x_0\theta(z) + \xi\vartheta(z)]^2 + [y_0\theta(z) + \eta\vartheta(z)]^2 \\ &= r\theta^2 + 2t\theta\vartheta + s\vartheta^2 \end{aligned} \tag{35}$$

We next substitute the expressions for U_1, V_1 , X_1, P_1 from eqs. (34), (35), (19) and (22) into eq. (32). For example, the term $\int(H_{11}U_1 + H_{12}V_1)X_1 g dz$ becomes

$$\begin{aligned} &\int[H_{11}(rG^2 + 2tGg + sg^2) + H_{12}(r\theta^2 + 2t\theta\vartheta + s\vartheta^2)](x_0G + \xi g)gdz \\ &= [\int(H_{11}G^3g + H_{12}\theta^2Gg)dz]rx_0 \\ &+[\int(H_{11}G^2g^2 + H_{12}\theta^2g^2)dz]r\xi \\ &+[\int(2H_{11}G^2g^2 + 2H_{12}\theta\vartheta Gg)dz]tx_0 \\ &+[\int(2H_{11}Gg^3 + 2H_{12}\theta\vartheta g^2)dz]t\xi \\ &+[\int(H_{11}g^3G + 2H_{12}\vartheta^2Gg)dz]sx_0 \\ &+[\int(H_{11}g^4 + H_{12}\vartheta^2g^2)dz]s\xi \end{aligned} \tag{36}$$

In a similar manner the other terms can also be calculated. The final result can be put in the form

$$\begin{aligned} X_3(z_1) &= [As + 2Bt + (C + D)r]\xi + [Bs + 2Ct + Er]x_0 \\ Y_3(z_1) &= [As + 2Bt + (C + D)r]\eta + [Bs + 2Ct + Er]y_0 \end{aligned} \tag{37}$$

where A, B, C, D, and E are known as the aberration coefficients and are defined by the relations

$$
\begin{aligned}
A &= \frac{2}{\vartheta(z_1)}\int_{z_0}^{z_1}[H_{11}g^4 + 2H_{12}g^2\vartheta^2 + H_{22}\vartheta^4]dz \\
B &= \frac{2}{\vartheta(z_1)}\int_{z_0}^{z_1}[H_{11}Gg^3 + H_{12}g\vartheta(g\theta + G\vartheta) + H_{22}\theta\vartheta^3]dz \\
C &= \frac{2}{\vartheta(z_1)}\int_{z_0}^{z_1}[H_{11}G^2g^2 + 2H_{12}Gg\theta\vartheta + H_{22}\theta^2\vartheta^2]dz \qquad (38) \\
D &= \frac{2}{\vartheta(z_1)}\Gamma^2\int_{z_0}^{z_1}H_{12}dz \\
E &= \frac{2}{\vartheta(z_1)}\int_{z_0}^{z_1}\left[H_{11}G^3g + H_{12}G\theta(g\theta + G\vartheta) + H_{22}\theta^3\vartheta\right]dz
\end{aligned}
$$

where Γ, called the paraxial optical invariant of the system is defined by

$$\Gamma = G\vartheta - g\theta$$

The coefficients A, B, C, D, and E are called the aberration coefficients, and their physical significance will be discussed in the next section. They depend only on the system parameters and the object and image planes.

Physical significance of the coefficients *A*, *B*, *C*, *D*, and *E*

In this section we study what kind of image pattern is produced separately by each of the terms containing A, B, C, D, and E. They correspond, respectively to the five Seidel aberrations: spherical aberration, coma, astigmatism, curvature of field, and distortion.

Since the system is rotationally symmetric, we can without loss of generality choose our object point to lie on the x axis. i.e., $y_0 = 0$, and the object would have coordinates $(x_0, 0, z_0)$ with some chosen origin of reference. We will study the image pattern produced by the system when all rays passing through the system are considered. Let $z = \zeta$ be the exit pupil plane. It is easier to consider the image formed due to sets of rays striking the exit pupil plane at a constant distance ρ from the z axis. Then this distance can be varied from zero to the radius of the exit pupil to get the complete set of rays forming the image.

The values of ξ and η for rays hitting the exit pupil plane $z = \zeta$ at a constant distance ρ from the z axis are given by (see Fig. 7.3)

$$\begin{aligned} \xi &= \rho \cos \varphi \\ \eta &= \rho \sin \varphi \end{aligned} \tag{38}$$

Different values of φ would correspond to the different rays striking the plane $z = \zeta$ along different points on the circle of radius ρ centered at $\xi = \eta = 0$. Since $y_0 = 0$ we have

$$\begin{aligned} r &= x_0^2 \\ t &= x_0 \rho \cos \varphi \\ s &= \rho^2 \end{aligned} \tag{39}$$

and the aberrations X_3 and Y_3 reduce to

$$\begin{aligned} X_3 &= A\rho^3 \cos \varphi + Bx_0\rho^2(2 + \cos 2\varphi) + (3C + D)x_0^2\rho \cos \varphi + Ex_0{}^3 \\ Y_3 &= A\rho^3 \sin \varphi + Bx_0\rho^2 \sin 2\varphi + (C + D)x_0^2\rho \sin \varphi \end{aligned} \tag{40}$$

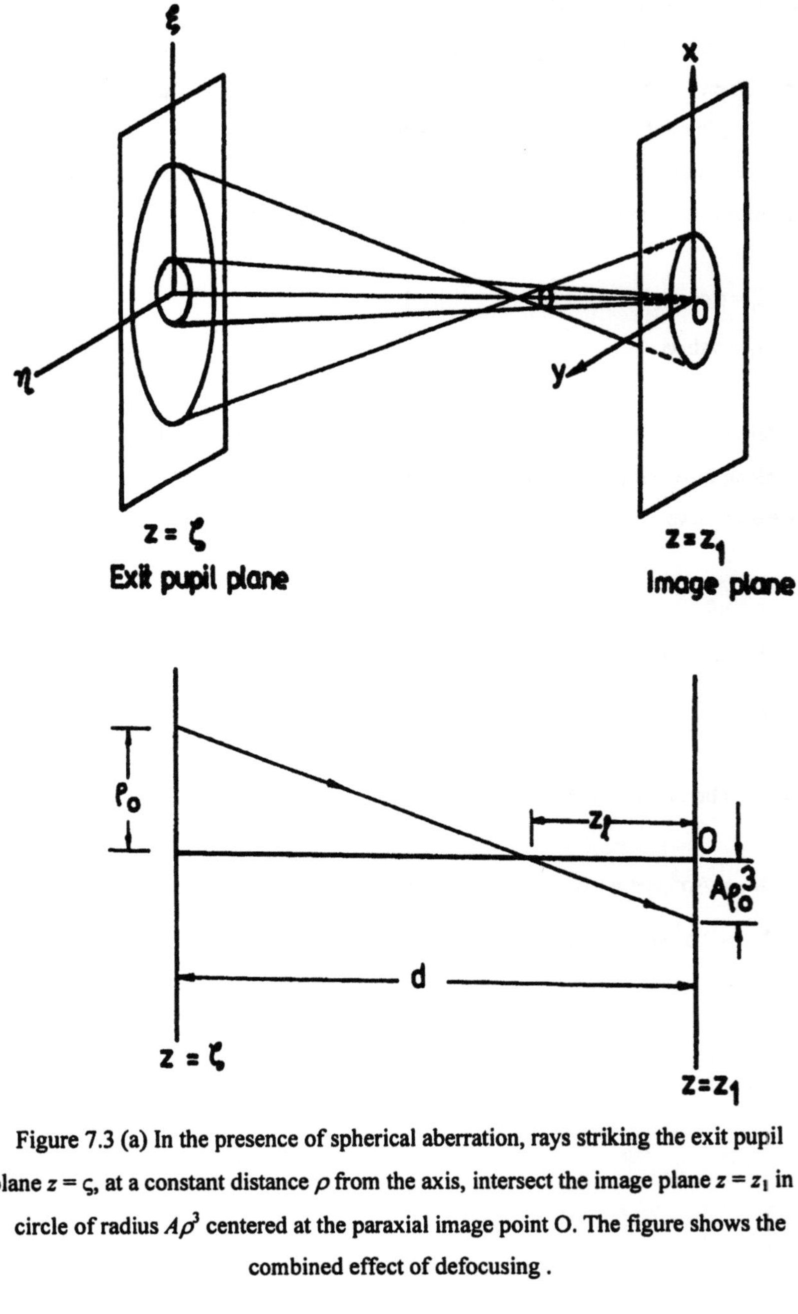

Figure 7.3 (a) In the presence of spherical aberration, rays striking the exit pupil plane $z = \varsigma$, at a constant distance ρ from the axis, intersect the image plane $z = z_1$ in a circle of radius $A\rho^3$ centered at the paraxial image point O. The figure shows the combined effect of defocusing .

Spherical aberration

If A is the only nonzero coefficient, then the aberrations along the x and y directions are

$$X_3 = A\rho^3 \cos\varphi$$
$$Y_3 = A\rho^3 \sin\varphi \quad (42)$$

It should be noted that even in the presence of other aberrations, for an axial object (i.e., $x_0 = 0$, $y_0 = 0$) X_3 and Y_3 are rigorously given by Eq. (42) and the other aberrations do not make any contribution. From Eq. (42) we get

$$(X_3^2 + Y_3^2) = (A\rho^3)^2 \quad (43)$$

X_3 and Y_3 represent the deviations of the points of intersection of the ray from the paraxial image point along the x and y directions, respectively. Consequently, rays that strike the exit pupil plane at a constant distance ρ from the axis intersect the paraxial image plane $z = z_1$ in a circle of radius $A\rho^3$ whose center lies at the paraxial image point O (see fig. 7.3a). Thus in the presence of this aberration the image of a point object is a circular patch of light. If the radius of the exit pupil is ρ_0, then the image would be a circular patch of light of radius $A\rho_0^3$. The distance $A\rho_0^3$ represents the lateral spherical aberration. Thus, when a set of rays parallel to the axis hits a lens, the rays striking the edge of the lens focus to a point called the marginal focus, which is different from the paraxial focus. To calculate the longitudinal spherical aberration (i.e., the aberration along the axis of the system) we refer to fig. 7.3(b) and find

$$\frac{A\rho^3}{\rho} = \frac{z_1}{(d - z_1)}$$

or

$$z_1 \approx A\rho^2 d \quad (44)$$

where z_l (known as the longitudinal spherical aberration) is the distance between the marginal focus and the paraxial focus and d is the distance between the exit pupil and the image plane. Thus the paraxial rays focus at the paraxial image plane and the rays coming from the edge of the exit aperture of radius ρ_0 focus at a distance z_l (= $A\rho_0^2 d$) from the paraxial image plane. When a screen is placed at the paraxial image plane then the image is a patch of light of radius $A\rho_0^3$. When the screen is placed at the marginal focus, the image is again a patch of light, due to the defocusing of the paraxial rays. Hence somewhere in between, one expects the circular patch to have a minimum diameter. This image is called the circle of least confusion (see fig. 7.3(a)).

Coma

As already mentioned, the only aberration present in the image for an on-axis object point is spherical aberration. The first off-axis aberration from which the image suffers is coma. This aberration is represented by B. For very small values of x_0, as can be seen from Eq. (41) the aberration in the image is due to A and B only. Let us consider the effect of the term containing B on the image of a point object. The aberrations are from Eq. (41)

$$\begin{aligned} X_3 &= Bx_0\rho^2(2+\cos 2\varphi) \\ Y_3 &= Bx_0\rho^2 \sin 2\varphi \end{aligned} \tag{45}$$

Eliminating φ from eq. (45), we get

$$(X_3 - 2Bx_0\rho^2)^2 + Y_3^2 = (Bx_0\rho^2)^2 \tag{46}$$

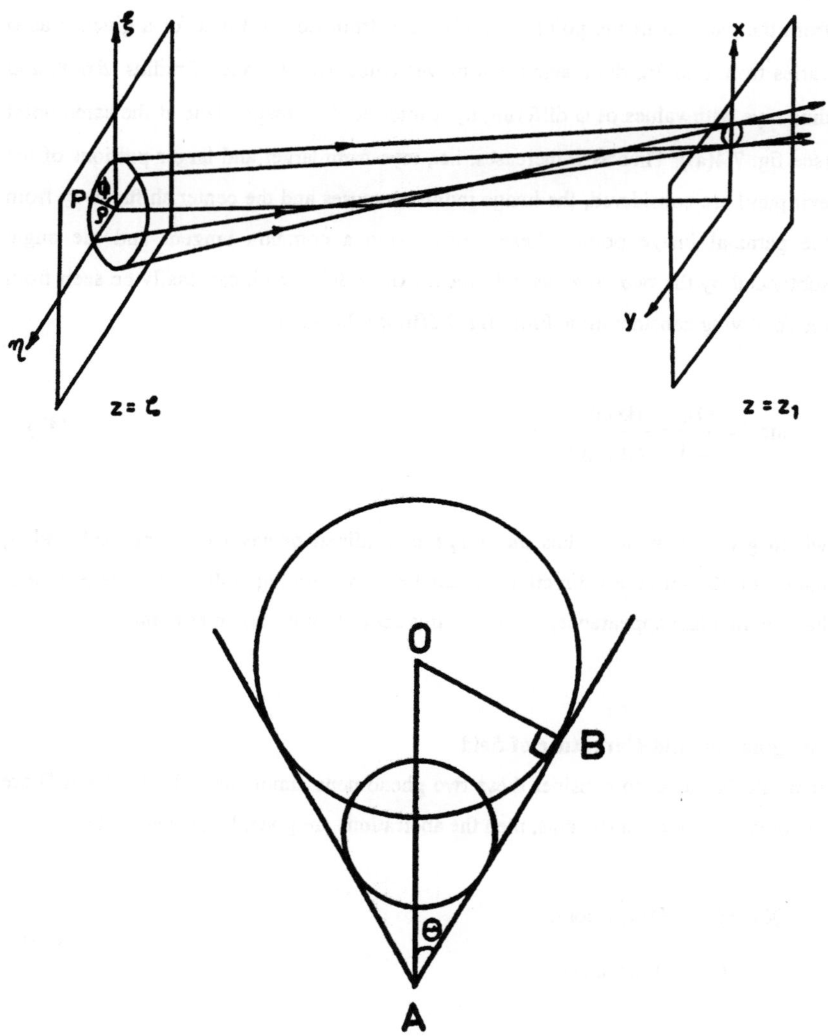

Figure 7.4 (a) In the presence of coma, the rays passing through the circumference of a circle of radius ρ on the exit pupil plane ($z = \varsigma$) intersect the image plane on a circle of radius $Bx_0\rho^2$ [OB (see Fig. b)] whose center is displaced from the paraxial image point A by $2Bx_0\rho^2$ [OA (see Fig. b)]. Notice that rays with the same value of ρ and the values of φ differing by π intersect the image plane at the same point. (b) The image of a point object afflicted with coma is confined within two lines subtending angles $\theta = 30^0$ with the x direction.

Equation (46) represents a circle of radius $Bx_0\rho^2$ whose center is at a distance $2Bx_0\rho^2$ from the paraxial image point. It can be seen from Eq. (45) that for a given ρ as φ varies from 0 to 2π, the image point moves twice over a circle of radius $Bx_0\rho^2$, and thus rays with values of φ differing by π intersect the image plane at the same point (see fig. 7.4(a)). Thus as ρ increases, i.e., rays from larger and larger portions of the exit pupil are considered, the image radius increases and the center shifts away from the paraxial image point. These circles have a common tangent and the angle subtended by the two tangents with the x axis is 30^0, which can easily be seen from the following consideration. From fig. 7.5(b) it follows that

$$\sin\theta = \frac{OA}{OB} = \frac{Bx_0\rho^2}{2Bx_0\rho^2} = \frac{1}{2} \tag{47}$$

which gives $\theta = 30^0$. Thus the image is confined between two lines subtending angles of 30^0 with the x direction. As can be seen from Fig. 7.4(b), the image pattern has a comet like appearance, and hence this aberration is known as coma.

Astigmatism and Curvature of field

It would be easier to consider these two phenomena simultaneously. If C and D are the only non-zero coefficients, then the aberrations are given by (see eq. (41))

$$\begin{aligned} X_3 &= (3C + D)x_0^2\rho\cos\varphi \\ Y_3 &= (C + D)x_0^2\rho\sin\varphi \end{aligned} \tag{48}$$

These equations can be combined to give

$$\frac{X_3^2}{[(3C + D)x_0^2\rho]^2} + \frac{Y_3^2}{[(C + D)x_0^2\rho]^2} = 1 \tag{49}$$

which is the equation of an ellipse with semi-axes $(3C + D)\, x_0^2\rho$ and $(C + D)\, x_0^2\rho$, with its center at the paraxial image point. Thus rays emanating from a given value of ρ intersect the paraxial image plane in an ellipse (see fig. 7.5(a)).

Let us consider a plane parallel to the paraxial image plane and specified by a coordinate z measured from the plane AB (see Fig. 7.5(b)). Let ($\bar{x}$, $\bar{y}$) represent the variables in this plane (see Fig. 7.5(b)). Let d be the distance between the exit pupil plane and the paraxial image plane. Then by consideration of similar triangles we can obtain*

$$\bar{x} = -\frac{z}{d}\xi + \left(1 + \frac{z}{d}\right)x_1 \tag{50}$$

where (x_1, y_1) represent the coordinates of the point of intersection of the ray with the paraxial image plane, and

$$x_1 = X_1 + (3C + D)x_0^2\xi \tag{51}$$

where the point (X_1, Y_1) represents the paraxial image point. Substituting for x_1 from eq. (51) into eq. (50), we get

$$\bar{x} = \frac{d+z}{d}X_1 + \left[-\frac{z}{d} + \frac{d+z}{d}(3C + D)x_0^2\right]\xi \tag{52}$$

Similarly, we can obtain

$$\bar{y} = \left[-\frac{z}{d} + \frac{d+z}{d}(C + D)x_0^2\right]\eta \tag{53}$$

where we have used the fact that $Y_1 = 0$ since the object lies in the $x - z$ plane. Since $\xi^2 + \eta^2 = \rho^2$, we have

---

* The space between the exit pupil and the image plane is assumed to be homogeneous.

$$\frac{[\bar{x}-(d+z)X_1/d]^2}{[-z/d+(d+z)(3C+D)x_0^2/d]^2}+\frac{\bar{y}^2}{[-z/d+(d+z)(C+D)x_0^2/d]^2}=\rho^2 \quad (54)$$

Equation (54) tells us that in a plane at a distance $-z$ from the paraxial image plane, a set of rays with a fixed ρ and varying φ form an ellipse centered at the point $\{[(d + z/d)]X_1, 0\}$. For certain values of z, the ellipse degenerate into a single line. When $z = z_T$ such that

$$-\frac{z_T}{d}+\frac{d+z_T}{d}(3C+D)x_0^2=0$$

or

$$z_T \cong (3C+D)x_0^2 d \quad (55)$$

then the ellipse degenerates into a straight line along the y direction (shown as a in fig. 7.5(a)). This is called the tangential focus. Similarly, when $z = z_s$ such that

$$-\frac{z_s}{d}+\frac{d+z_s}{d}(C+D)x_0^2=0$$

or

$$z_s \approx (C+D)x_0^2 d \quad (56)$$

the ellipse degenerate into a straight line along the x direction (shown as e in fig. 7.5(a)). This is called the sagittal focus. The distance between these two foci, namely,

$$z_T - z_s = 2Cx_0^2 d \quad (57)$$

is a measure of astigmatism. Thus the two foci would coincide when $C = 0$.

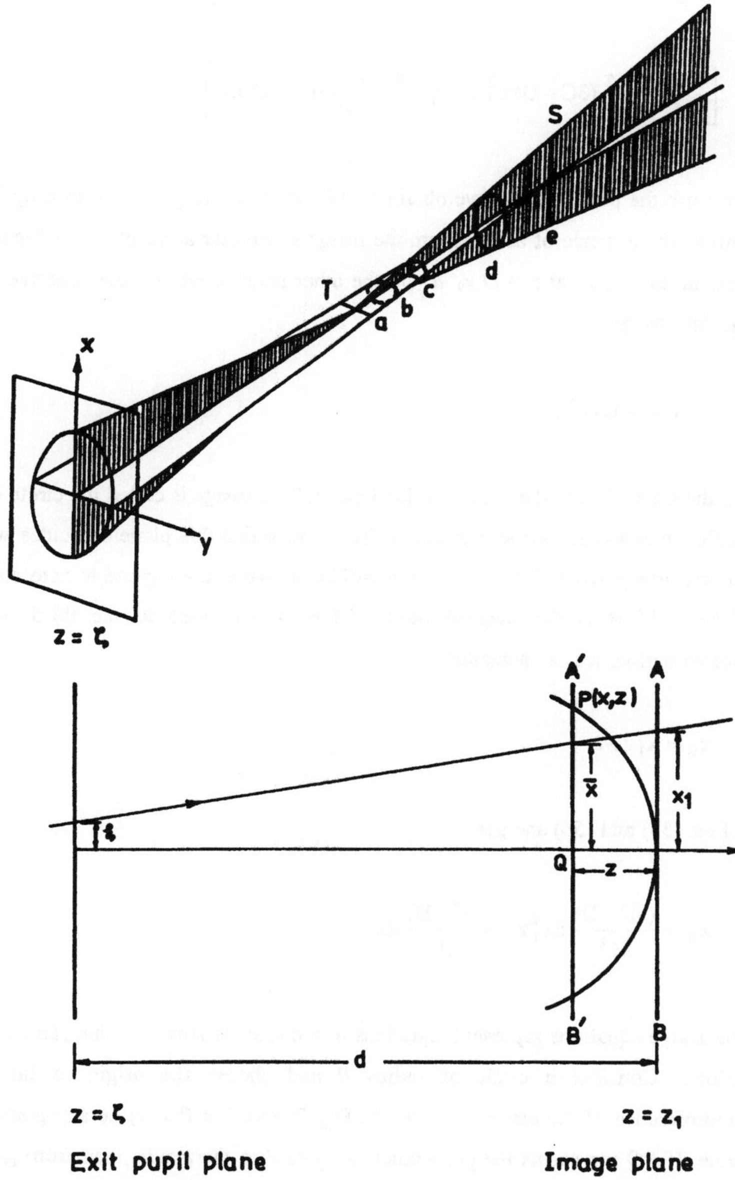

Figure 7.5 (a) Focusing of rays in the presence of astigmatism. *a* represents the tangential focus and *e* the sagittal focus. (b) *AB* represents the paraxial image plane and *A′B′* any other nearby plane parallel to *AB*, at a distance *z* from it.

The ellipse degenerates into a circle when

$$\left[-\frac{z}{d}+\frac{d+z}{d}(3C+D)x_0^2\right]=\pm\left[-\frac{z}{d}+\frac{d+z}{d}(C+D)x_0^2\right] \tag{58}$$

If we use the positive sign, we obtain $C = 0$, which corresponds to no astigmatism. Thus in the absence of astigmatism the image is circular at all planes, which would generate to a point at $z = D\, x_0^2\, d$. On the other hand, if we use the negative sign in Eq. (58), we get

$$z=(2C+D)x_0^2 d$$

i.e., the image is circular only at a fixed plane. This image is called the circle of least confusion and is shown as *c* in Fig. 7.5(a). Notice that this plane coincides with the paraxial image when $C = -D/2$, which will be shown to correspond to zero curvature of field. If M is the magnification of the system, then to the third order of approximation, we can substitute

$$x_0 = x_1 / M \tag{59}$$

in Eqs. (55) and (56) and get

$$z_T=\frac{(3C+D)}{M^2}dx_1^2 z_s=\frac{(C+D)}{M^2}dx_1^2 \tag{60}$$

The above equations represent equations of a circle for small x_1; this can be seen as follows. Consider a circle of radius R and choose the origin to lie on the circumference of the circle as shown in Fig. 7.5(b). Let $P(x, z)$ be any point on the circle. If PQ represents the perpendicular from P to the z axis, then from geometry one obtains rigorously $-z\,(-2R+z) = x^2$, the solution of which (that goes to 0 as $x \to 0$) is

$$z = R - R\left(1 - \frac{x^2}{R^2}\right)^{1/2} \approx \frac{x^2}{2R} \tag{61}$$

Thus the curves given by Eq. (60) represent circles of radii R_T and R_S given by

$$\begin{aligned} & z_\Gamma \approx (3C+D)x_0^2 \\ & \frac{1}{R_T} = \frac{2d}{M^2}(3C+D)\quad \frac{1}{R_S} = \frac{2d}{M^2}(C+D) \end{aligned} \tag{62}$$

These are called the tangential and sagittal field curvatures, respectively. The mean of these two,

$$\frac{1}{R} = \frac{1}{2}\left(\frac{1}{R_T} + \frac{1}{R_S}\right) = \frac{2d}{M^2}(2C+D) \tag{63}$$

is called the field curvature.

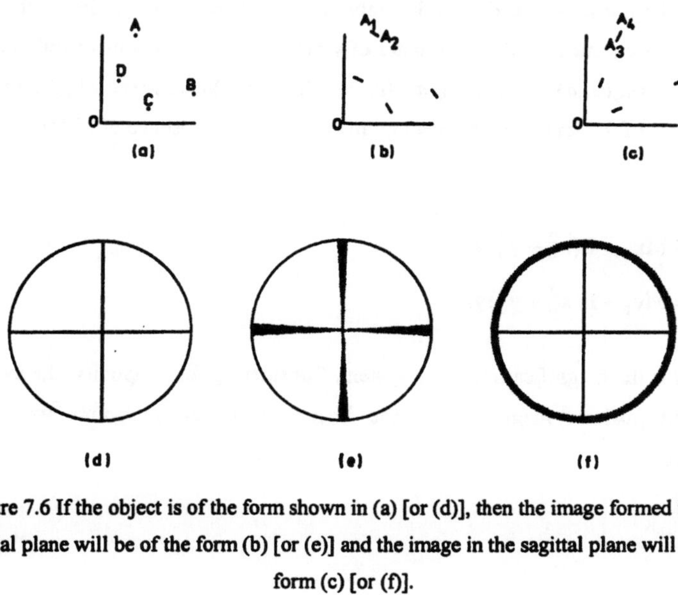

Figure 7.6 If the object is of the form shown in (a) [or (d)], then the image formed in the tangential plane will be of the form (b) [or (e)] and the image in the sagittal plane will be of the form (c) [or (f)].

Let us consider four object points *A, B, C* and *D* in the object plane (see Fig. 7.6(a)). In the tangential plane, each point will be imaged as a straight line, which will be perpendicular to the line joining the object point and the axial point *O*. For example, the point *A* will be imaged along a line A_1A_2, which is in a direction perpendicular to *OA*. Similarly, in the sagittal plane the point *A* will be imaged along a line A_3A_4, which is in a direction parallel to *OA*. Thus, if we have a spoked wheel in the object plane (see Fig. 7.6(d)) then the image in the tangential and sagittal planes will be similar to that shown in Figs. 7.6(e) and 7.6(f).

Distortion

The coefficient *E* represents the distortion in the image. In order to study its physical significance, we consider an object point specified by the coordinates (x_0, y_0) in the plane $z = z_0$. Neglecting the effect of other aberrations (i.e., assuming $A = B = C = D = 0$). Eq. (37) becomes

$$X_3 = E(x_0^2 + y_0^2)x_0 Y_3 = E(x_0^2 + y_0^2)y_0 \tag{64}$$

Let us consider a square grid as shown in Fig. 7.7(a). Let *a* represent the spacing between the lines. We need consider only one of the quadrants in detail; the others follow from symmetry. The coordinates of various intersection points would be (0,0), (*a*, 0), (2*a*, 0), (0, *a*), (0, 2*a*), (*a, a*), (*a*, 2*a*), (2*a, a*), (2*a*, 2*a*) (see Fig. 7.7(a)). The coordinates of the image of an object point (x_0, y_0) in the presence of distortion would be

$$\begin{aligned} x_1 &= Mx_0 + E(x_0^2 + y_0^2)x_0 \\ y_1 &= My_0 + E(x_0^2 + y_0^2)y_0 \end{aligned} \tag{65}$$

where *M* is the magnification of the system. Considering for simplicity, the system to have unit positive magnification, the image of the above points would have coordinates

$$0,0)(a + Ea^3 0)(2a + 4Ea^3 0)(0, a + Ea^3)(02a + 4Ea^3),$$

$$(a + 2Ea^3 a + 2Ea^3)(a + 5Ea^3 2a + 10Ea^3),$$

$$(2a + 10Ea^3 a + 5Ea^3)(2a + 16Ea^3 2a + 16Ea^3)$$

These have been plotted in Figs. 7.7(b) and 7.7(c) for positive and negative E, respectively. The dashed lines show the ideal image. It follows at once how positive E modifies the image into a form called pincushion distortion and negative E into a form called barrel distortion.

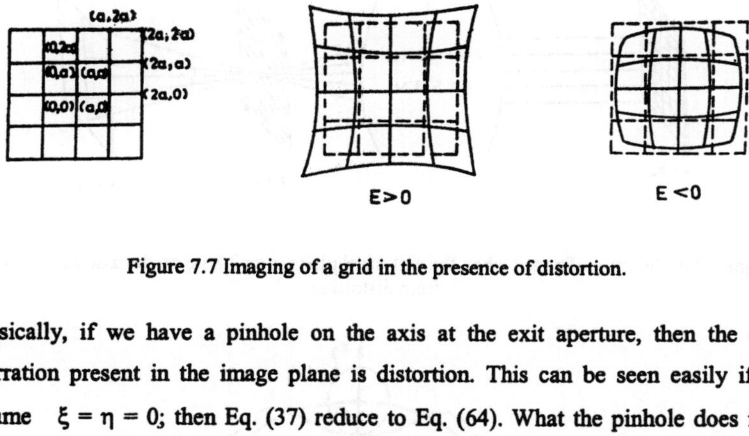

Figure 7.7 Imaging of a grid in the presence of distortion.

Physically, if we have a pinhole on the axis at the exit aperture, then the only aberration present in the image plane is distortion. This can be seen easily if we assume $\xi = \eta = 0$; then Eq. (37) reduce to Eq. (64). What the pinhole does is to allow only those rays that satisfy $\xi = \eta = 0$ (see Fig. 7.8). Further, when there is a pinhole, then corresponding to a point on the object plane only one ray will reach the image plane. Consequently there cannot be any aberration present except distortion of the object due to nonuniform magnification.

We now consider the case when all the third-order aberrations are present simultaneously. The resultant image shape will be a combination of all the aberration image shapes we have so far considered.

Equations (41) can be combined to yield

$$\{X_3 - 2Bx_0\rho^2 - Ex_0^3 - [A\rho^3 + (3C+D)x_0^2\rho]\cos\varphi\}^2 + \{Y_3 - [A\rho^3 + (C+D)x_0^2\rho]\sin\varphi\}^2 = (Bx_0\rho^2)^2 \qquad (66)$$

Equation (66) is the equation of a circle of radius $Bx_0\rho^2$ whose center lies on an ellipse with axes $A\rho^3 + (3C + D)\, x_0^2\rho$ and $A\rho^3 + (C + D)\, x_0^2\rho$. The center of the ellipse in turn lies at $(2B\, x_0\rho^2 + Ex_0^3, 0)$. The composite shape is shown in Fig. 7.9. Every time the center of the circle moves around the ellipse once, any point on the circumference of the circle moves around twice.

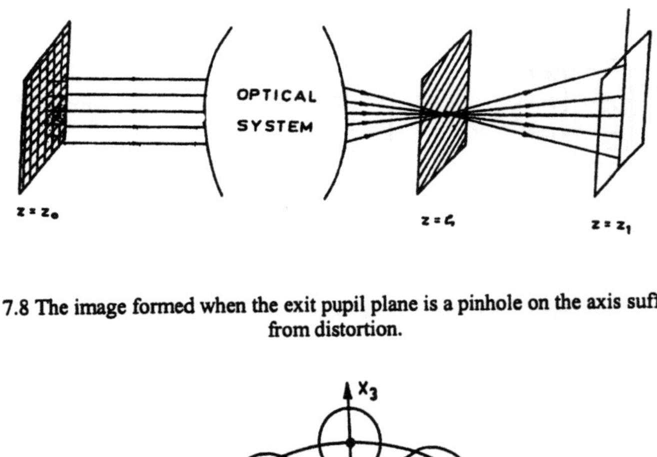

Figure 7.8 The image formed when the exit pupil plane is a pinhole on the axis suffers only from distortion.

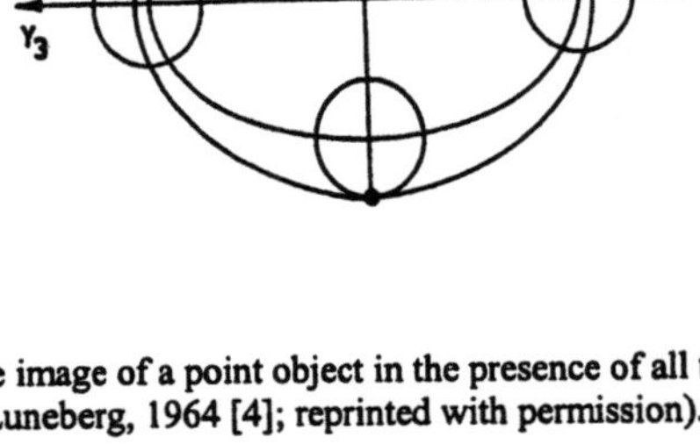

Figure7.9 Form of the image of a point object in the presence of all the aberrations (after Luneberg, 1964 [4]; reprinted with permission).

The coefficients H_{ij} in terms of refractive index variation

Since the series on the right-hand side of Eq. (7) is essentially a Taylor series expansion of the Hamiltonian H around $U = 0$, $V = 0$, we may write the coefficients H_1, H_2, etc., in the following form:

$$H_1 = \left.\frac{\partial H}{\partial U}\right|_{U=0,V=0}, \quad H_2 = \left.\frac{\partial H}{\partial V}\right|_{U=0,V=0}$$

$$H_{11} = \left.\frac{\partial^2 H}{\partial U^2}\right|_{U=0,V=0}, \quad H_{12} = \left.\frac{\partial^2 H}{\partial U \partial V}\right|_{U=0,V=0}, \quad H_{22} = \left.\frac{\partial^2 H}{\partial V^2}\right|_{U=0,V=0} \tag{67}$$

Using the expression $H = -[n^2(U, z)-V]^{1/2}$, we obtain

$$\begin{aligned} H_1 &= \left.\frac{\partial [n^2(U,z)-V]^{1/2}}{\partial U}\right|_{U=0,V=0} \\ &= -\left.\frac{n(U,z)}{[n^2(U,z)-V]^{1/2}}\frac{\partial n(U,z)}{\partial U}\right|_{U=0,V=0} = -\left.\frac{\partial n(U,z)}{\partial U}\right|_{U=0} \end{aligned} \tag{68}$$

Similarly

$$H_2 = \frac{1}{2n(0,z)}, \qquad H_{11} = -\left.\frac{\partial^2 n(U,z)}{\partial U^2}\right|_{U=0}$$

$$H_{12} = -\frac{1}{2n^2(0,z)}\left.\frac{\partial n(U,z)}{\partial U}\right|_{U=0}, \quad H_{22} = \frac{1}{4n^3(0,z)}, \tag{69}$$

Thus given an index distribution $n(U, z)$, the above coefficients can be calculated. Using the values of H_1 and H_2, the paraxial equations have to be solved under specific boundary conditions to obtain the axial and the field rays. Once they have been obtained, the explicit values for the aberration coefficients can be calculated easily.

In Section 'Aberration of graded-index media', we make an explicit calculation of the third-order aberration in a continuously varying inhomogeneous medium. In Section 'Aberration in systems possessing finite discontinuities in refractive index' the aberrations of systems consisting of lenses formed of homogeneous media will be considered. The general case of systems consisting of inhomogeneous media separated by surfaces has been discussed elsewhere [e.g.,5].

Aberrations of graded-index media

In this section we will calculate the aberrations in a medium characterized by the following refractive index variation:

$$n(U,z) = n_0\left(1-\frac{1}{2}\alpha^2 U+\frac{1}{2}\beta\alpha^4 U^2\right) \tag{70}$$

A general rotationally symmetric inhomogeneous medium contains higher order terms also, like U^3, U^4. However, we need not consider these higher order terms since they would be important only for fifth and higher order aberrations.* This is because of the fact that the coefficients of the cubic and higher powers of U and V in the expansion of the Hamiltonian H [see Eq. (7)] do not contribute to third-order aberration. Such media exhibit focussing properties (see Chapter 6) and can be used in the form of image relays and as microlenses of ultrashort focal lengths. As such the study of aberrations introduced by such media is important [7]. These fibers can also be used in optical communications.

Although n_0, α and β may in general be continuous functions† of z, we consider them to be independent of z, which is almost true in actual practical fibers. For such a refractive-index distribution,

---

* An analysis of fifth-order aberrations of inhomogeneous media has been carried out in Ref. 6.

† Conical Selfoc fibers are characterized by refractive index variation of the form

$$H_1 = \frac{n_0\alpha^2}{2} H_2 = \frac{1}{2n_0}$$
$$H_{11} = -n_0\beta\alpha^4 H_{12} = \frac{\alpha^2}{4n_0} H_{22} = \frac{1}{4n_0^3} \tag{71}$$

and hence the paraxial equations [Eqs. (12) and (14)], from which the axial and field rays may be derived, reduce to

$$\frac{dX_1}{dz} = \frac{1}{n_0} P_1 \frac{dP_1}{dz} = -n_0\alpha^2 X_1 \tag{72}$$

The general solution of Eq. (72) when n_0 and α are independent of z is

$$X_1 = A\sin\alpha z + B\cos\alpha z$$
$$P_1 = n_0\alpha(A\cos\alpha z - B\sin\alpha z) \tag{73}$$

We choose the plane $z = \zeta$ such that

$$\sin\alpha\zeta = 1 \tag{74}$$

Under such a condition, the axial (g, ϑ) and the field (G, θ) rays are given by

$$g(z) = \sin\alpha z \vartheta(z) = n_0\alpha\cos\alpha z$$
$$G(z) = \cos\alpha z \theta(z) = -n_0\alpha\sin\alpha z \tag{75}$$

where the object plane has been chosen to be $z = 0$. Hence we have a series of paraxial image planes at values of z satisfying

$$\sin\alpha z_1 = 0 \tag{76}$$

$$n(U, z) = n_0[1 - \frac{1}{2}\alpha^2\kappa(z)U + \frac{1}{2}\beta\alpha^4U^2]$$

where $\kappa(z) = (1 - \sigma z)^{-2}$, σ being a constant. Third-order aberrations of such media have been obtained in Ref.7.

i.e., $\alpha z_1 = m\pi$, $m = 1,2,3,\ldots$. Choosing $m = 2$ for simplicity, we have

$$z_1 = 2\pi/\alpha \tag{77}$$

Having obtained g, ϑ, G, θ, H_{11}, H_{12}, H_{22}, z_0, and z_1, we have to substitute these in Eqs. (38) to get explicit values of A, B, C, D, and E. For example, the coefficient of spherical aberration is given by

$$\begin{aligned} A = \frac{2}{n_0\alpha}\int_0^{z_1}\Big[& -n_0\beta\alpha^4 \sin\alpha z + \frac{\alpha^2}{2n_0} n_0^2\alpha^2 \cos^2\alpha z \sin^2\alpha z \\ & + \frac{1}{4n_0^3} n_0^4 \alpha^4 \cos^2\alpha z \Big] dz \\ = {} & z_1\alpha^3\left(\frac{5}{16} - \frac{3}{4}\beta\right) \end{aligned} \tag{78}$$

where use has been made of the fact that if

$$I_n \equiv \int_0^{z_1} (\sin\alpha z)^n (\cos\alpha z)^{4-n}\, dz \tag{79}$$

then

$$I_0 = \frac{3z_1}{8}, I_1 = 0, I_2 = \frac{z_1}{8}, I_3 = 0, I_4 = \frac{3z_1}{8} \tag{80}$$

Similarly

$$B = 0, C = -\frac{z_1\alpha^3}{4}(\beta + \frac{1}{4}), D = \frac{1}{2}z_1\alpha^3, E = 0 \tag{81}$$

Thus the system is free from coma ($B = 0$) and distortion ($E = 0$). The total aberrations along the x and y directions can now be written as [see Eq. (37)]

$$X_3 = \left(\frac{5}{16} - \frac{3}{4}\beta\right) z_1 \alpha^3 \xi (x_0^2 + \xi^2 + \eta^2)$$

$$Y_3 = z_1 \alpha^3 \left[\left(\frac{7}{16} - \frac{1}{4}\beta\right) x_0{}^2 + \left(\frac{5}{16} - \frac{3}{4}\beta\right)(\xi^2 + \eta^2)\right]\eta \qquad (82)$$

where, as before, the object point is specified by $(x_0, 0)$.

Let us first consider the case when the image is formed by meridional rays only. These rays satisfy the condition that they always remain in a plane containing the optical axis, which in the present case is the z axis. Since the object point lies in the x-z plane, the condition for a ray to be meridional would be $\eta = 0$. In such a case, from Eq. (82) we get

$$X_3 = \left(\frac{5}{16} - \frac{3}{4}\beta\right) z_1 \alpha^3 \xi (x_0^2 + \xi^2), Y_3 = 0 \qquad (83)$$

It can be seen from Eq. (83) that when β = 5/12, the third-order aberrations in meridional rays can be made identically zero. In such a case the index distribution given by Eq. (70) reduces to

$$n(U) = n_0 (1 - \frac{1}{2}\alpha^2 U + \frac{5}{24}\alpha^4 U^2) \qquad (84)$$

$$n(U) = n_0 (1 - \frac{1}{2}\alpha^2 U + \frac{3}{8}\alpha^4 U^2)$$

This can be seen to be first three terms of an index function

$$n(U) = n_0 \operatorname{sech}(\alpha U^{1/2}) \qquad (85)$$

We have here been able to prove that such an index distribution is free from third-order aberrations as far as meridional rays are concerned, but it can be proved by other detailed considerations that such an index distribution is indeed free from all orders of aberrations as far as meridional rays are concerned. Meridional rays are not the only type of rays that take part in image formation. Of course, if the object point happened to be axial, all rays emanating from the object would be meridional and such an index distribution would produce ideal images. This can be seen from the fact (as already observed in Section on Physical significance of coefficients A,B,C,D and E) that the only type of aberration present for an axial object is the spherical aberration, and for an index distribution given by Eq. (85) the spherical aberration is identically zero.

Rays that do not remain in one plane containing the optical axis are called skew rays. A special kind of these are helical rays, which maintain a constant distance from the optical axis as they travel through the medium; they trace a helix and hence the name (see Fig. 6.5b). For such helical rays (see Chapter 6)

$$\begin{aligned} X_1(z) &= x_0 \cos \alpha z \\ Y_1(z) &= x_0 \sin \alpha z \end{aligned} \tag{86}$$

At the reference plane $z = \zeta = \pi/2\,\alpha$ and $X_1(\zeta) = \xi = 0$ and $Y_1(\zeta) = \eta = x_0$. Thus using Eq. (82), we obtain

$$X_3 = 0,\ Y_3 = z_1 \alpha^3 \eta^3 \left(\frac{3}{4} - \beta\right) \tag{87}$$

Hence for helical rays to be free from third-order aberrations we must have $\beta = 3/4$. Under such a condition the refractive-index distribution given by Eq. (70) becomes

$$n(U) = n_0 \left(1 - \frac{1}{2}\alpha^2 U + \frac{3}{8}\alpha^4 U^2\right) \tag{88}$$

which can be seen to represent the first three terms of

$$n(U) = n_0 / (1 + \alpha^2 U)^{1/2} \tag{89}$$

This refractive index distribution can also be derived from Fermat's principle, and it can be shown that for such an index variation the helical rays travel without aberrations of any order (see Chapter 4).

Thus it is clear from the above analysis how certain specific index distributions are free from aberrations for certain specific subclasses of rays, meridional and helical. Since all kinds of rays, including skew rays, are involved in image formation, we can see that any particular index distribution cannot be completely free from aberrations for all kinds of rays. In the above analysis we have not considered any surfaces to be present in the system. These would also add to the aberrations. Here lies the importance of inhomogeneous media, because the aberrations can be controlled by changing the parameters of the inhomogeneous medium in a system.

Aberrations in systems possessing finite discontinuities in refractive index

In Section on Aberration of graded-index media, we considered the aberrations of media possessing continuous variation of refractive index with position. In this section we modify the formulas for *A*, *B*, *C*, *D*, and *E* into a form suitable for direct application to systems composed of lenses.

We will first derive formulas for a single surface formed between two homogeneous media. This will later be extended to systems consisting of any number of surfaces.

The refractive index variation of a system composed of two homogeneous media of refractive indices n_1 and n_2 separated by a surface with a radius of curvature R is [see Eq. (18) of Chapter 6]

$$n(U,z) = n_1\Theta(\Lambda) + n_2\Theta(-\Lambda) \tag{90}$$

where $\Lambda = f(U) - z$, $U = X^2 + Y^2$, $\Theta(x)$ is the unit step function and

$$z = f(U) \simeq \frac{U}{2R} + \frac{U^2}{8R^3} \tag{91}$$

represents the equation of the spherical surface. It immediately follows from Eq. (90) that for $z < f(U)$, $n = n_1$, and for $z > f(U)$, $n = n_2$.

For evaluating the quantities H_i and H_{ij} we have to determine the first and second derivatives of H with respect to U. Thus from Eq. (90) we

$$\begin{aligned}\frac{dn}{dU} &= n_1 \frac{d\Theta(\Lambda)}{d\Lambda}\frac{df}{dU} - n_2 \frac{d\Theta(-\Lambda)}{d(-\Lambda)}\frac{df}{dU} \\ &= -\Delta n_1 \delta(\Lambda)\frac{df}{dU}\end{aligned} \tag{92}$$

where use has been made of Eq. (21) of Chapter 6 and $\Delta n_1 = n_2 - n_1$. Differentiating again, we obtain

$$\frac{d^2 n}{dU^2} = -\Delta n_1 \delta(\Lambda)\frac{d^2 f}{dU^2} - \Delta n_1 \frac{d\delta(\Lambda)}{d\Lambda}\left(\frac{df}{dU}\right)^2 \tag{93}$$

Thus, using Eqs. (92) and (93) in Eqs. (68) and (69) we get

$$\begin{aligned}&H_1 = \frac{\Delta n_1}{2R}\delta(z), H_2 = \frac{1}{2n_0(z)}, H_{12} = \frac{\Delta n_1}{4n_0^2(z)R}\delta(z) \\ &H_{11} = \frac{\Delta n_1}{4R^3}\delta(z) - \frac{\Delta n_1}{4R^2}\delta'(z), H_{22} = \frac{1}{4n_0^3(z)}\end{aligned} \tag{94}$$

where the prime denotes differentiation with respect to the argument and we have used the fact that $\delta(-z) = \delta(z)$. Further, $n_0(z) = n_1\,\Theta(-z) + n_2\,\Theta(z)$, which represents the refractive-index variation along the z-axis. If we substitute for H_{11}, H_{12}, and H_{22} in Eq. (38), we will obtain the expressions for the various aberration coefficients. For example,

$$A = \frac{1}{2\vartheta(z_1)} \int_{z_0}^{z_1} \left\{ \left[\delta(z) g^4 - R\delta'(z) g^4 + 2\frac{R^2}{n_0^2}\delta(z) g^2 \vartheta^2 \right] \frac{\Delta n_1}{R^3} + \frac{\vartheta^4}{n_0^3} \right\} dz$$

$$= \frac{1}{2\vartheta(z_1)} \int_{z_0}^{z_1} \left\{ \left[g^4 + 4Rg^3 \frac{dg}{dz} + 2\frac{R^2}{n_0^2} g^2 \vartheta^2 \right] \frac{\Delta n_1}{R^3} \delta(z) + \frac{\vartheta^4}{n_0^3} \right\} dz \quad (95)$$

where we have used the following property of the delta function:

$$\int \delta'(z) f(z) dz = -\int f'(z) \delta(z) dz \quad (96)$$

Since

$$\frac{dg}{dz} = 2H_2\vartheta = \frac{1}{n_0}\vartheta \quad (97)$$

we obtain

$$\int_{z_0}^{z_1} \frac{1}{2n_0^3} \vartheta^4 dz = \int_{z_0}^{z_1} \frac{1}{2n_0^2} \vartheta^3 \frac{dg}{dz} dz$$

$$= \frac{\vartheta^3}{2n_0^2}\Bigg|_{z_0}^{z_1} - \int_{z_0}^{z_1} \left\{ \frac{1}{2n_0^2} 3\vartheta^2 \left(-\frac{\Delta n_1}{R} \delta(z) g \right) \right.$$

$$\left. - \frac{\vartheta^3}{n_0^3} \Delta n_1 \delta(z) \right\} g dz \quad (98)$$

$$= \int_{z_0}^{z_1} \left(\frac{3g^2\vartheta^2}{2n_0^2 R} + g\frac{\vartheta^3}{n_0^3} \right) \dot{n}_0 dz$$

where we have used

$$\dot{n}_0 \equiv \frac{dn_0}{dz} = \frac{d}{dz}[n_1\Theta(-z) + n_2\Theta(z)] = \Delta n_1 \delta(z) \tag{99}$$

$$\frac{d\vartheta}{dz} = -2H_1 g = -\frac{\Delta n_1}{R}\delta(z)g = -\dot{n}_0 \frac{g}{R} \tag{100}$$

and the fact that

$$g(z_0) = g(z_1) = 0 \tag{101}$$

(see Section on 'Expressions for third order aberration'). Using Eqs. (98) and (100) in Eq. (95), we obtain

$$\vartheta(z_1)A = \frac{1}{2}\int_{z_0}^{z_1}\left(\frac{g^3}{R^3} + 4\frac{g^2}{R^2}\frac{\vartheta}{n_0} + 5\frac{g}{R}\frac{\vartheta^2}{{n_0}^2} + 2\frac{\vartheta^3}{{n_0}^3}\right)\dot{n}_0 g dz \tag{102}$$

One can use Eqs. (97), (99) and (100) to prove that

$$\begin{aligned}&\left(\frac{g^3}{R^3} + 4\frac{g^2}{R^2}\frac{\vartheta}{n_0} + 5\frac{g}{R}\frac{\vartheta^2}{{n_0}^2} + 2\frac{\vartheta^3}{{n_0}^3}\right)\dot{n}_0 \\ &= -\left[\frac{d}{dz}\left(\frac{\vartheta}{n_0}\right)\right]^2\left[\frac{d}{dz}\left(\frac{\vartheta}{{n_0}^2}\right)\right]\Big/\left[\frac{d}{dz}\left(\frac{1}{n_0}\right)\right]^2\end{aligned} \tag{103}$$

Thus Eq. (102) reduces to

$$\vartheta(z_1)A = -\frac{1}{2}\int_{z0}^{z_1} g\left\{\left[\frac{d}{dz}\left(\frac{\vartheta}{n_0}\right)\right]\left[\frac{d}{dz}\left(\frac{\vartheta}{{n_0}^2}\right)\right]\Big/\left[\frac{d}{dz}\left(\frac{1}{n_0}\right)\right]^2\right\}dz \tag{104}$$

The above formula can be extended to systems consisting of any number of surfaces. Between two points in a homogeneous medium, n_0 and ϑ remain constant and the integrand in Eq. (104) vanishes. Thus the contribution to A from a homogeneous medium is zero. Consequently, the above integral may be evaluated around the discontinuous only. For a discontinuity at $z = 0$, the integral should be evaluated from $z = -\varepsilon$ to $z = \varepsilon$. Now, for an arbitrary function $G_1(z)$ that is discontinuous at $z = 0$, we may write (around $z = 0$)

$$dG_1 / dz = \delta(z)\Delta G_1 \tag{105}$$

where ΔG_1 represents the change in G_1 at the discontinuity. Since ϑ and n_0 have a discontinuity at $z = 0$, we immediately obtain

$$\vartheta(z_1)A = -\frac{1}{2}g_s\left\{\left[\Delta\left(\frac{\vartheta_1}{n_1}\right)\right]^2\left[\Delta\left(\frac{\vartheta_1}{n_1^{\,2}}\right)\right]\Big/\left[\Delta\left(\frac{1}{n_1}\right)\right]^2\right. \tag{106}$$

where g_s is the x coordinate of the axial ray on the surface and as before ΔX represents the change in X on refraction, i.e.,

$$\left[\Delta\left(\frac{1}{n_1}\right)\right] = \left(\frac{1}{n_2} - \frac{1}{n_1}\right) = \frac{1}{n_1}(k-1) \tag{107}$$

$$\Delta\left(\frac{\vartheta_1}{n_1}\right) = \left(\frac{\vartheta_2}{n_2} - \frac{\vartheta_1}{n_1}\right) \tag{108}$$

where ϑ_1 and ϑ_2 are the values of ϑ before and after refraction and $k = n_1/n_2$. Integrating Eq. (100) from a point to the left of the surface to a point to the right of the surface, we have

$$\vartheta_2 - \vartheta_1 = -\frac{\Delta n_1}{2R}g_s \tag{109}$$

Using Eq. (109), Eq. (108) reduces to

$$\Delta\left(\frac{\vartheta_1}{n_1}\right) = (k-1)\left(\frac{g_s}{R} + \frac{\vartheta_1}{n_1}\right) \tag{110}$$

We also have

$$\Delta\left(\frac{\vartheta_1}{n_1^2}\right) = \left(\frac{\vartheta_2}{n_2^2} - \frac{\vartheta_1}{n_1^2}\right) = \frac{(k-1)}{n_2}\left(\frac{g_s}{R} + \frac{\vartheta_1}{n_1} + \frac{\vartheta_1}{n_1 k}\right) \tag{111}$$

Using Eq. (107), (110), and (111), Eq. (106) reduces to

$$A_s = -\frac{1}{2\vartheta(z_1)} n_1 (k-1) g_s \left(\frac{g_s}{R} + \frac{\vartheta_1}{n_1}\right)^2 \left(\frac{g_s}{R} + \frac{\vartheta_1}{n_1} + \frac{\vartheta_2}{n_2}\right) \tag{112}$$

This is the contribution to A from each surface of the system, where the quantities are referred to their values on the surface. Similarly we can obtain

$$\begin{aligned} &B_s = qA_s, D_s = -\frac{1}{2\vartheta(z_1)}\Gamma^2 \frac{1}{R}\Delta\left(\frac{1}{n_1}\right) \\ &C_s = q^2 A_s, E_s = q^3 A_s + qD_s \end{aligned} \tag{113}$$

where

$$q = \left(\frac{G_s}{R} + \frac{\theta_1}{n_1}\right) \Big/ \left(\frac{g_s}{R} + \frac{\vartheta_1}{n_1}\right) \tag{114}$$

G and θ represent the coordinates of the field ray. Hence to calculate aberrations one has to first trace the axial and field rays and then calculate the values of A_s , B_s , C_s, D_s and E_s due to each surface, and then add the contribution due to each surface to obtain the complete aberration present in the system. We will now consider some examples.

A plane glass surface

Consider the image formed by a plane glass surface, of an object O as shown in Fig. 7.10. The image is virtual but the analysis that we have developed remains valid. The reference plane $z = \zeta$ is the plane of the glass surface and the relative index of the refraction of the two media is μ; the point I represents the paraxial image point (see Fig. 7.10). To find the axial ray, consider a paraxial ray starting from O and hitting the plane $z = \zeta = 0$ at a unit height. Hence

$$g_s = g(z=\zeta) = 1, \vartheta = 1 \sin\alpha_1 \simeq \alpha_1 \simeq -1/z_0 \qquad (115)$$

$$\vartheta' = \mu \sin\alpha_2 \simeq \mu\alpha_2 \simeq -\mu/z_1 \qquad (116)$$

Observe that both z_0 and z_1 are negative quantities. Since the rays satisfy Snell's law at the boundary ($z = \zeta$), we have $\sin\alpha_1 = \mu \sin\alpha_2$, which for paraxial rays gives $z_1 = \mu z_0$, which gives the position of the paraxial image point. Thus, in the paraxial approximation, from Eq. (116) we obtain

$$\vartheta' = -1/z_0 \qquad (117)$$

Substituting these values in the expression for the coefficient of spherical aberration [see Eq. (112)], we obtain

$$\begin{aligned} A &= -\frac{1}{2(-1/z_0)} 1 \times \left(\frac{1}{\mu} - 1\right) \times 1 \times \left(0 - \frac{1}{z_0}\right)^2 \times \left(0 - \frac{1}{z_0} - \frac{1}{\mu z_0}\right) \\ &= \frac{\mu^2 - 1}{2 z_0^2 \mu^2} \end{aligned} \qquad (118)$$

where we have used the fact that for a plane glass surface $R = \infty$. Thus for a ray striking the surface at a height l from the z axis, the lateral spherical aberration is

$$X_3 = Al^3 = \frac{\mu^2 - 1}{2 z_0^2 \mu^2} l^3 \qquad (119)$$

The longitudinal spherical aberration Δz is given by [see Eq. (44)]

$$\Delta z = Al^2 z_1 = \frac{\mu^2 - 1}{2 z_0 \mu} l^2 \tag{120}$$

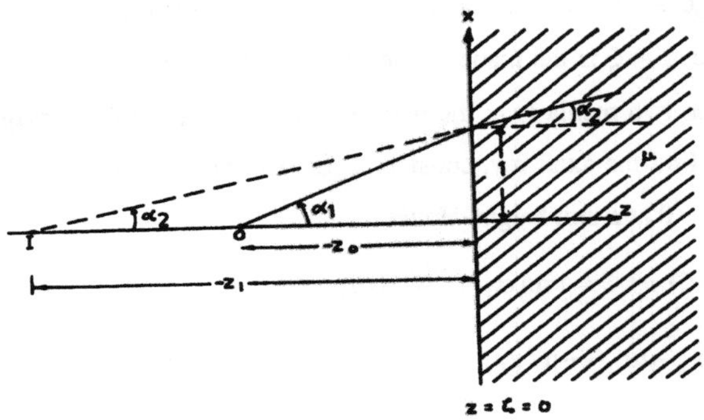

Fig. 7.10 Aberration introduced by a plane refracting surface separating two media of relative refractive index μ. The surface is taken to be the exit pupil plane.

Aberration of a thin lens

We will now obtain explicit expressions for the various aberrations present in a thin lens. To determine the aberrations of a thin lens we have first to trace the axial and field rays. Let z_0 be the distance of the object plane from the lens (since the lens is assumed to be thin, z_0 is the distance from either the first or the second surface) and let the distance of the paraxial image plane from the lens (see Fig. 7.11) be z_1. Let R_1 and R_2 be the radii of curvatures of the two surfaces and let μ be the refractive index of the medium of the lens. Let g_0, g_1, g_2 and g_i represent the height of the axial ray in the object plane, the first and second refracting surfaces, and the image plane, respectively; clearly

$$g_0 = 0, g_1 = 1, g_2 = 1, g_i = 0 \tag{121}$$

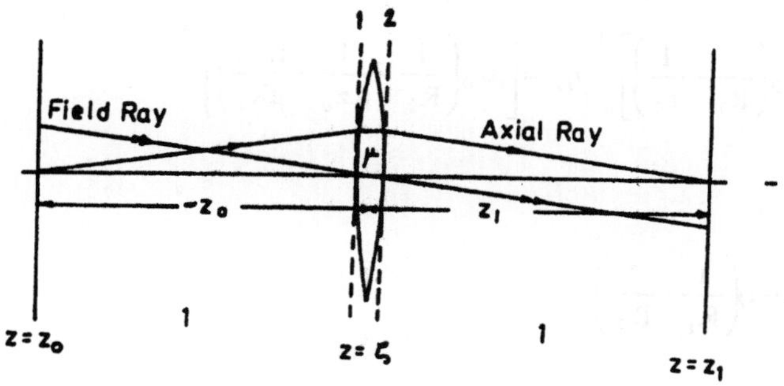

Figure 7.11 The configuration for calculating the aberrations of a thin lens. The axial ray and the field ray are also shown.

Further using the analysis of Section on Thick lens in Chapter 6, we obtain the following expressions for the optical direction cosines in the three media:

$$\vartheta_1 = -\frac{1}{z_0} \tag{122}$$

$$\vartheta_2 = \vartheta_1 - \frac{1}{R_1}(\mu - 1) = -\frac{1}{z_0} - \frac{\mu - 1}{R_1} \tag{123}$$

$$\vartheta_3 = \vartheta_2 + \frac{\mu - 1}{R_2} = -\frac{1}{z_0} - (\mu - 1)\left(\frac{1}{R_1} - \frac{1}{R_2}\right) = -\frac{1}{z_1} \tag{124}$$

Similarly, for the field ray we obtain

$$G_0 = 1, G_1 = 0, G_2 = 0, G_i = -z_1 / z_0 \tag{125}$$

and

$$\theta_1 = 1/z_0, \theta_2 = 1/z_0, \theta_3 = 1/z_0 \tag{126}$$

Further

$$q_1 = \left[z_0 \left(\frac{1}{R_1} - \frac{1}{z_0} \right) \right]^{-1}, q_2 = \left[\mu z_0 \left(\frac{1}{R_2} - \frac{1}{\mu z_0} - \frac{\mu - 1}{\mu R_1} \right) \right]^{-1} \tag{127}$$

$$\wp = (\mu - 1)\left(\frac{1}{R_1} - \frac{1}{R_2} \right) \tag{128}$$

Using the above expressions , we can evaluate the aberration coefficients *A*,*B*,*C*,*D* and *E*. For example, the contribution from the first surface to the spherical aberration is [see Eq. (112)]

$$\begin{aligned} A_{s1} = & -\frac{1}{2(-1/z_1)} 1 \times \left(\frac{1}{\mu} - 1 \right) \times 1 \times \left(\frac{1}{R_1} + \frac{1}{z_0} \right) \\ & \times \left[\frac{1}{R_1} + \frac{1}{z_0} + \frac{1}{\mu z_0} - \frac{(\mu - 1)}{\mu} \frac{1}{R_1} \right] \\ = & \frac{(\mu - 1) z_1}{2\mu} \left(\frac{1}{R_1} - \frac{1}{z_0} \right)^2 \left(\frac{\mu + 1}{\mu z_0} - \frac{1}{\mu R_1} \right) \end{aligned} \tag{129}$$

The contribution from the second surface is

$$\begin{aligned} A_{s2} = & \frac{z_1 \mu}{2} (\mu - 1) \left(\frac{1}{R_2} + \frac{1}{\mu} \left(\frac{1}{z_0} - \frac{\mu - 1}{R_1} \right) \right)^2 \\ & \times \left[\frac{1}{R_1} + \frac{1}{\mu} \left(\frac{1}{z_0} - \frac{(\mu - 1)}{R_1} \right) - \frac{1}{z_1} \right] \\ = & \frac{\mu(\mu - 1) z_1}{2} \left\{ \left[\frac{1}{\mu} \left(-\frac{1}{z_0} + \frac{1}{R_1} \right) - \frac{\wp}{\mu - 1} \right]^3 \right. \\ & \left. - \frac{1}{z_1} \left[\frac{1}{\mu} \left(-\frac{1}{z_0} + \frac{1}{R_1} \right) - \frac{\wp}{\mu - 1} \right]^2 \right\} \end{aligned} \tag{130}$$

Thus the coefficient of spherical aberration A for the complete lens is the sum of Eq. (129) and (130), which after some algebraic simplifications can be put in the form

$$
\begin{aligned}
A &= A_{s1} + A_{s2} \\
&= \frac{z_1(\mu-1)}{2\mu^2}\left\{\left[\left(\frac{1}{R_2}-\wp\right)^2\left(\frac{1}{R_2}-\wp(\mu+1)\right)-\frac{1}{R_1^3}\right]\right. \\
&\quad -\frac{1}{z_0}\left[(\mu+3)\left(\frac{1}{R_2^2}-\frac{1}{R_1^2}+\wp^2-\frac{2\wp}{R_2}\right)-2\mu\wp\left(\frac{1}{R_2}-\wp\right)\right] \\
&\quad \left.+\frac{1}{z_0^2}\left[(2\mu+3)\left(\frac{1}{R_2}-\frac{1}{R_1}-\wp\right)-\mu\wp\right]\right\}
\end{aligned}
\tag{131}
$$

Similarly the other aberration coefficients can be obtained

$$
B=\frac{z_1(\mu-1)}{2(-z_0)}\left[\frac{(\mu-1)(2\mu+1)}{\mu R_1 R_2}-\frac{\mu^2-\mu-1}{\mu^2 R_1^2}-\frac{\mu}{R_2^2}\right]+\frac{(2\mu+1)}{z_0^2\mu(\mu-1)}\wp \tag{132}
$$

$$
C=\frac{z_1\wp}{2z_0^2}, D=-\frac{z_1\wp}{2\mu z_0^2}, E=0 \tag{133}
$$

It is interesting to note that the thin lens is free from distortion ($E = 0$). A simple case of this analysis is when the object is at infinity. Then for rays incident parallel to the axes one can use these formulas by taking z_0 to $-\infty$. When $z_0 \rightarrow -\infty$, as is evident B, C, and D also go to zero and A reduces to

$$
A=\frac{f(\mu-1)}{2\mu^2}\left[\left(\frac{1}{R_2}-\wp\right)^2\left(\frac{1}{R_2}-\wp(\mu+1)\right)-\frac{1}{R_1^3}\right] \tag{134}
$$

In this case the image is at the focus, and the aberration determines the difference in paraxial and marginal foci. It should be borne in mind that the aberration obtained by the above calculation is the lateral spherical aberration, i.e., the aberration in a plane normal to the axis. Thus the lateral spherical aberration is

$$S_{lat} = -\frac{(\mu-1)f}{2\mu^2}\left[\frac{1}{R_1^{\,3}} - \left(\frac{1}{R_2} - \frac{\mu+1}{f}\right)\left(\frac{1}{R_2} - \frac{1}{f}\right)^2\right]\rho^3 \tag{135}$$

where f is the focal length of the thin lens $(=1/\wp)$ and ρ is the distance from the axis at which the ray strikes the lens. From here one can easily obtain the longitudinal spherical aberration, i.e., the difference between the paraxial and marginal foci [see Eq. (44); in the present case $d = f$]:

$$S_{long} = -\frac{(\mu-1)f^2\rho^2}{2\mu^2}\left[\frac{1}{R_1^{\,3}} - \left(\frac{1}{R_2} - \frac{\mu+1}{f}\right)\left(\frac{1}{R_2} - \frac{1}{f}\right)^2\right] \tag{136}$$

This represents the difference in foci between rays striking the lens at a distance ρ from the axis and the paraxial rays.

The image formed by a set of rays parallel to the axis of the lens suffers only from spherical aberration. If we consider a set of parallel rays inclined to the axis, then the image suffers from other aberrations also. To calculate the coma in an image formed by a set of parallel rays inclined to the axis at an angle φ, we consider an object at a height x_0, from the axis at a distance $-z_0$ from the lens. Then the coma present in the image is $3Bx_0\rho^2$. Using Eq. (132) this becomes

$$\begin{aligned}\text{coma} = {} & \frac{3(\mu-1)\cdot}{2} z_1 \left(\frac{x_0}{-z_0}\right)\rho^2\left[\frac{(\mu-1)(2\mu+1)}{\mu R_1 R_2} - \frac{\mu^2-\mu-1}{\mu^2 R_1^2} - \frac{\mu}{R_2^2}\right] \\ & + \frac{3(2\mu+1)\wp z_1}{\mu(\mu-1)}\rho^2\left(\frac{x_0}{z_0^2}\right)\end{aligned} \tag{137}$$

In the limit of $x_0 \to -\infty$ and $z_0 \to -\infty$ such that $x_0/z_0 = \tan\varphi$ remains constant, we get a parallel set of rays inclined at an angle φ to the axis. Thus the amount of coma in the image formed by a parallel bundle of rays incline at an angle φ to the z axis is given by

$$\text{coma} = \frac{3(\mu-1)}{2} f\rho^2 \tan\varphi \left[\frac{(\mu-1)(2\mu+1)}{\mu R_1 R_2} - \frac{\mu^2-\mu-1}{\mu^2 R_1^2} - \frac{\mu}{R_2^2}\right] \tag{138}$$

We consider a thin lens made up of a material of refractive index 1.5. Let the radii of curvatures of the two surfaces be –40 and –8 cm. We next calculate the longitudinal spherical aberration of a ray incident parallel to the axis at a height 1 cm from the axis. We also calculate the coma when a ray is incident at a height 1 cm from the axis at an angle $\varphi = \tan^{-1}(1/5)$.

The focal length of the lens can be calculated to be f = 20 cm. Then substituting the given values of the parameters in Eq. (136) and (138), we obtain

$$S_{long} = -0.34\text{cm}, \text{coma} = -0.03\text{cm}$$

As another example, we consider a thin biconvex lens with the following values of the parameters: R_1 = 20 cm, R_2 = -20 cm, n = 1.5. Consider an object point situated at a distance 30 cm from the lens and at a height 1 cm from the axis. We will determine the different amounts of aberrations present in an image formed by rays striking the lens at a height of 1 cm.

The focal length of the lens can be calculated to be f = 20 cm. Hence the power $\wp$ = $1/f$ = 0.05 cm $^{-1}$. The image point lies at a distance of 60 cm from the lens. Then using Eqs. (131)-(133) one can obtain

$$A = -8.89\times10^{-3}\text{cm}^{-2}, B = -3.11\times10^{-3}\text{cm}^{-2}$$

$$C = 1.67\times10^{-3}\text{cm}^{-2}, D = -1.11\times10^{-3}\text{cm}^{-2}, E = 0$$

Thus, the longitudinal spherical aberration $A\rho^2 z_1$ = 0.5334 cm; the coma $3Bx_0\rho^2$ = -0.009318 cm, and the distance between sagittal and tangential foci $-2Cx^2{}_0 z_1$ = - 0.2

cm. Further, the radius of curvature of the field $M^2/2d(2C+D) = 14.9$ cm, where $M = z_1/(-z_0)$ is the magnification of the system.

In general, for a combination of lenses , one must first trace out the axial and field rays and find the values of $g(z)$ and $G(z)$ at various refracting surfaces. The values of ϑ and θ in various media are also determined, from which one can obtain the aberration coefficients.

Chromatic aberration

The refractive index of any material medium depends on the wavelength of the radiation under consideration. Since image formation is accompanied by refraction at refractive-index discontinuities or by propagation through continuously varying refractive index media, the property of dispersion affects image formation. If the light is not monochromatic, then after refraction, different wavelength components proceed along different directions and hence form images at different points. This introduces aberration in the image and this aberration is called chromatic aberration. Since white light is normally used for image formations in most optical instruments, this aberration leads to colored images.

In this section we study the primary chromatic aberration of an optical system, where we assume the absence of any of the five Seidel aberrations (the five monochromatic aberrations discussed in sections on 'Expressions for third order aberration' and 'Physical significance of coefficients A,B,C,D and E'). The difference between the actual image point and the image point for any specified wavelength is the chromatic aberration. The component of this difference along the optical axis is called longitudinal chromatic aberration and the component normal to the optical axis is called lateral chromatic aberration.

For studying chromatic aberration, we must consider only the paraxial equations. The paraxial equations are

$$\dot{x} = 2H_2 p, \dot{p} = -2H_1 x \tag{139}$$

Let $z = z_0$ be the object plane with respect to some conveniently chosen origin. Since we are going to study the effect of wavelength on the image, we have to choose a particular wavelength and define aberrations with respect to the image formed by light of this wavelength. We define our axial ray (coordinates g, ϑ) and field ray (coordinates G, θ) with respect to this wavelength. (It is normally chosen to be λ = 5893 A). Let $z = z_1$ be the paraxial image plane with respect to this wavelength.

We will first obtain the chromatic aberration of a single refracting surface. It will be shown that the chromatic aberration for a system of surfaces can be readily obtained from the formula for a single refracting surface. Let R be the radius of curvature of the refracting surface separating two media of refractive indices n_1 and n_2. In such a case, as in Section ' Aberrations in system possessing finite discontinuities in refractive index' the refractive index function for the system would be

$$n(u,z) = n_1\Theta(\Lambda - z) + n_2\Theta(z - \Lambda) \tag{140}$$

where $\Lambda = f(u) + z_2$ and z_2 is the pole of the surface. As before $u = x^2 + y^2$ and $\Theta(x)$ is the unit step function. The equation of the surface is [see Eq. (17) of Chapter 6]

$$z = z_2 + f(u) \simeq z_2 + \frac{u}{2R} + \frac{u^2}{8R^3} + \ldots \tag{141}$$

The coefficients H_1 and H_2 have already been determined in Section 7.6:

$$H_1 = \frac{1}{2R}\dot{n}_0, H_2 = \frac{1}{2n_0} \tag{142}$$

where dots represent differentiation with respect to z and n_0 is the refractive index variation along the optical axis at the chosen reference wavelength. Thus Eq. (139) becomes

$$\dot{x} = \frac{1}{n_0} p, \dot{p} = -\frac{1}{R} \dot{n}_0 x \tag{143}$$

If $\delta(x)$ represents the change in x for a change $\delta\lambda$ in λ, then from Eq. (143) we obtain

$$\delta\dot{x} = \frac{1}{n_0} \delta p + p\delta\left(\frac{1}{n_0}\right), \delta\dot{p} = -\frac{1}{R}(\dot{n}_0 \delta x + x\delta\dot{n}_0) \tag{144}$$

Since (g, ϑ) is a solution of the paraxial equations, we have as before

$$\dot{g} = \frac{1}{n_0} \vartheta, \dot{\vartheta} = -\frac{1}{R} \dot{n}_0 g \tag{145}$$

We substitute the values of $1/n_0$ and $\dot{n}_0 /R$ in Eq. (144) to obtain

$$\frac{d}{dz}(\vartheta\delta x - g\delta p) = p\vartheta\delta\left(\frac{1}{n_0}\right) + \frac{xg}{R}\delta\dot{n}_0 \tag{146}$$

Integrating Eq. (146) from z_0 to z_1, where $z = z_0$ and $z = z_1$ represent the object and the image planes, respectively, and using the fact that $g(z_0) = g(z_1) = 0$ and $(\delta x)_{z_0}$ (which represents the chromatic aberration in the object plane) is zero, we have

$$\vartheta(z_1)(\delta x)_{z_1} = \int_{z_0}^{z_1} \left[p\vartheta\delta\left(\frac{1}{n_0}\right) + \frac{xg}{R}\delta\dot{n}_0 \right] dz \tag{147}$$

Using

$$x \equiv x_0 G(z) + \xi\, g(z), p = x_0 \theta(z) + \xi\vartheta(z) \tag{148}$$

where x_0 is the x coordinate of the ray in the object plane $z = z_0$, and ξ is the x coordinate in the exit pupil plane $z = \zeta$, Eq. (147) reduces to

$$(\delta x)_{z_1} = K\xi + Lx_0 \tag{149}$$

where

$$K = \frac{1}{\vartheta(z_1)} = \int_{z_0}^{z_1} \left[\vartheta^2 \delta\left(\frac{1}{n_0}\right) + \frac{g^2}{R}\delta \dot{n}_0 \right] dz$$
$$L = \frac{1}{\vartheta(z_1)} = \int_{z_0}^{z_1} \left[\theta\vartheta\delta\left(\frac{1}{n_0}\right) + \frac{gG}{R}\delta \dot{n}_0 \right] dz \tag{150}$$

$$K = \frac{1}{\vartheta(z_1)} = \int_{z_0}^{z_1} \left[\vartheta^2 \delta\left(\frac{1}{n_0}\right) + \frac{g^2}{R}\delta \dot{n}_0 \right] dz$$
$$L = \frac{1}{\vartheta(z_1)} = \int_{z_0}^{z_1} \left[\theta\vartheta\delta\left(\frac{1}{n_0}\right) + \frac{gG}{R}\delta \dot{n}_0 \right] dz$$

We can simplify Eqs. (150) by using Eq. (145). For example,

$$K = \frac{1}{\vartheta(z_1)} \int_{z_0}^{z_1} \left[-\frac{\vartheta}{n_0}\delta n_0 \frac{dg}{dz} - g\frac{\dot{\vartheta}}{\dot{n}_0}\delta \dot{n}_0 \right] dz \tag{151}$$

Integrating the first term in Eq. (151) by parts and using the fact that $g(z_0) = g(z_1) = 0$, we obtain

$$K=\frac{1}{\vartheta(z_1)}\int_{z_0}^{z_1}\left[-\frac{\dot{\vartheta}}{\dot{n}_0}\delta\dot{n}_0+\frac{\dot{\vartheta}}{n_0}\delta n_0+\frac{\vartheta}{\dot{n}_0}\delta\dot{n}_0-\frac{\vartheta}{n_0{}^2}\dot{n}_0\delta n_0\right]dz$$

$$=\frac{1}{\vartheta(z_1)}\int_{z_0}^{z_1}g\left[\frac{d}{dz}\left(\frac{\vartheta}{n_0}\right)\frac{d}{dz}\left(\frac{\delta n_0}{n_0}\right)\Big/\frac{d}{dz}\left(\frac{1}{n_0}\right)\right]dz \tag{152}$$

Similarly,

$$L=\frac{1}{\vartheta(z_1)}\int_{z_0}^{z_1}g\left[\frac{d}{dz}\left(\frac{\theta}{n_0}\right)\frac{d}{dz}\left(\frac{\delta n_0}{n_0}\right)\Big/\frac{d}{dz}\left(\frac{1}{n_0}\right)\right]d \tag{153}$$

where $\delta n_0 / n_0$ measures the dispersion of the medium. Dispersion is normally measured by the quantity

$$\frac{n_F - n_C}{n_D - 1}$$

where n_F, n_C and n_D represent the refractive indices of the medium at wavelengths corresponding to the hydrogen *F* line (λ = 4861 A), hydrogen *C* line (λ = 6563 A), and sodium D line (λ = 5893 A), respectively. For typical values of n_F, n_C and n_D for various common optical materials, can be found in standard handbooks of physics and chemistry.

For an optical system composed of homogeneous media separated by surfaces, the contribution to *K* and *L* is only at the surfaces, and the chromatic aberration for the complete system is given by Eq. (149) where

$$K=\frac{1}{\vartheta(z_1)}\sum_i g_i\left[\Delta_i\left(\frac{\vartheta}{n}\right)\right]\frac{\Delta_i\kappa}{\Delta_i(1/n)}$$

$$L=\frac{1}{\vartheta(z_1)}\sum_i g_i\left[\Delta_i\left(\frac{\theta}{n}\right)\right]\frac{\Delta_i\kappa}{\Delta_i(1/n)} \tag{154}$$

The summation being done over the surfaces of the system. Here $\kappa=\delta n/n$.

The normal procedure employed is to remove the chromatic aberration for the F and C lines with respect to the D line. Even after this correction, the system would still be suffering from chromatic aberration for other wavelengths.

As an example we calculate the chromatic aberration of a thin lens mage up of a medium of refractive index n and placed in air. Let R_1 and R_2 represent the radii of curvature of the thin lens. We have already traced the axial and field rays through a thin lens (see Section on 'Aberrations in system possessing finite discontinuities in refractive index'). Thus the expressions for K and L as given by Eq. (154) can be obtained for a thin lens. For example, we have

$$K=-z_1\left\{1\times\left[\frac{1}{n}\left(-\frac{1}{z_0}-\frac{n-1}{R_1}\right)+\frac{1}{z_0}\right]\frac{\kappa}{(1/n)-1}\right.$$

$$\left.+1\times\left[-\frac{1}{z_1}-\frac{1}{n}\left(-\frac{1}{z_0}-\frac{n-1}{R_1}\right)\right]\frac{-\kappa}{(1/n)-1}\right\} \tag{155}$$

where the symbols have the same meaning as in Section "Aberrations in systems possessing finite discontinuities in refractive index". The total K is obtained as a sum of the contribution from each surface of the lens. On simplification, we obtain

$$K=-z_1\wp\frac{\delta n}{n-1} \tag{156}$$

In an exactly similar manner,

$$L=-z_1\left[1\times\left(\frac{1}{z_0 n}-\frac{1}{z_0}\right)\frac{\kappa}{(1/n)-1}+1\times\left(\frac{1}{z_0}-\frac{1}{z_0 n}\right)\frac{(-\kappa)}{1-(1/n)}\right]=0 \tag{157}$$

Thus, the lateral chromatic aberration is given by [see Eq. (149)]

$$\delta x = K\xi = -\wp z_1 \frac{\delta n}{n-1}\xi \tag{158}$$

where ξ is the height at which the ray strikes the exit pupil, which is the lens itself in the present case.

The longitudinal chromatic aberration (δz) is given by

$$\delta z = \frac{z_1}{\xi}(\delta x) = -\wp z_1^2 \frac{\delta n}{n-1} \tag{159}$$

For a parallel beam of light incident on the lens, $z_1 = f$, and the chromatic aberration is

$$\delta f = -f\frac{\delta n}{n-1} \tag{160}$$

It may be noted that for the simple case of a thin lens, the expression for chromatic aberration, namely, Eq. (160), could have been obtained in an alternative way as given below. The focal length of a thin lens is given by

$$\frac{1}{f} = (n-1)\left(\frac{1}{R_1} - \frac{1}{R_2}\right) \tag{161}$$

If a change of n by δn (the change of n is due to the change in the wavelength of the light) results in a change of f by δf, then by differentiating Eq. (161) we obtain

$$-\frac{\delta f}{f^2} = \delta n\left(\frac{1}{R_1} - \frac{1}{R_2}\right) = \frac{\delta n}{n-1}\frac{1}{f}$$

i.e.,

$$\delta f = -f\frac{\delta n}{n-1} \tag{162}$$

which is the same as Eq. (160).

We have dealt with aberrations in great detail in this chapter. In chapter 8 we introduce an operator technique based on lie algebras. This is because the field of adaptive optics has had an explosive growth in the past decade or so and hence warrants attention [8,9]. Adaptive optics utilizes real time correction of an aberrated wavefront to create a (nearly) aberration free image close to the diffraction limit. Knowing the aberrations it is possible to reconstruct the wavefront [e.g.,10]. Aberration compensation can be done using a number of methods such as deformable mirrors, liquid crystal spatial light modulators, microelectro machined component, membrane mirrors etc. to generate a "good" image. Adaptive optics is used for example, to compensate for atmospheric turbulence for ground based telescope imaging, underwater imaging where turbulence and turbidity limit imaging, optical line of sight communications, confocal microscopy, etc. One of the greatest advances in the use of adaptive optics is in vision research and ophthalmology. The reader is referred to the literature for details on this exciting area of optics research and technology.

References:

1. K. Thyagarajan and A K Ghatak, ***Hamiltonian Theory of third order aberrations of inhomogeneous lenses***, Optik **44**, 329-342, (1976)
2. V. N. Mahajan, ***Aberration theory made simple***, SPIE Optical Engineering Press, Bellingham, WA (1991)
3. W. T. Welford, ***Aberrations of optical systems***, Adam Hilger, Bristol, UK (1986)
4. R. Luneberg, ***Mathematical theory of optics***, University of California Press, Berkeley, (1964)
5. E. G. Rawson, D. R. Herriot, J. M. Mc Kenna, ***Analysis of refractive index distribution in cylindrical graded-index glass rods (GRIN Rods) used as image relays***, Applied Optics, **9**,753, (1970)
6. A. Gupta, K. Thyagarajan, I. C. Goyal, A. K. Ghatak, ***Theory of fifth order aberration of graded index media***. J. Opt. Soc. Am., **66**,1320-1326, (1976)
7. K. Thyagarajan, A. Rohra, A K Ghatak, ***Aberrations in conical graded index media***, Optica Acta, **26**: 863-871, (1979)
8. R. K. Tyson ed., ***Adaptive Optics Engineering Handbook***, Marcel Dekker, NY (1999)
9. R. K. Tyson. ***Principles of Adaptive Optics***, Second edition, Academic Press, San Diego, CA (1998)
10. V. Lakshminarayanan and J. B. Almeida, Wavefronts and Spherical aberration, in ***Vision Science and its Applications***, ed. V. Lakshminarayanan, Optical Society of America, Washington DC, Pages 178-186, (2000)
11. D. Miller, Adaptive optics in retinal microscopy in vision, in ***Handbook of Optics***, Vol . III., ed. M.Bass, et al., McGraw Hill, NY, Chapter 10, (2000)

8. AN INTRODUCTION TO LIE ALGEBRAIC TREATMENT OF OPTICAL ABERRATIONS

We have seen in the previous chapter on optical aberrations how complicated the algebra can get. Recently a number of workers [1-4] have introduced formalisms using matrix methods to do nonlinear calculations (i.e., departures for Gaussian optics) and hence calculate aberrations up to any arbitrary order. However, the drawback of matrix methods when extended to the non-linear regime is that in general, large matrices are required. For example, if one were to go up to the 5th order, the size of the matrices are of the dimensions of 125×125 for any arbitrary conical surface. The size of the matrices can be reduced by the use of symmetries, but the method is still cumbersome.

A generalized operator technique would be of great use, especially when considering any general asymmetrical optical system. Such a technique has been pioneered by Dragt and co-workers[5,6,7], using lie algebraic tools for characterizing optical systems. The fundamental idea is that geometrical optics can be formulated in terms of a variational principle involving a Lagrangian and from this one may find the equivalent Hamiltonian equations of motion as shown previously. If the Hamiltonian is known, then using the standard methods of Lie algebra, it is possible to describe light optics in terms of symplectic maps. For a general treatment of symplectic methods see the book by Guillemin and Sternberg [8]. Stavroudis has dealt extensively with group theoretical methods in optics [9]. These methods provide an operator extension of the linear matrix method to the general case and is the subject of this chapter. The reader is referred to the literature for detailed expositions.

Formalism and Mathematical background

Consider a general optical system. Assume that a ray originates at the initial point p^i with spatial coordinates $\vec{r}^i$ and moves in a direction specified by unit vector $\vec{s}^i$. After passing through the system, it comes to a point p^f with coordinates $\vec{r}^f$ in a direction specified by unit vector $\vec{s}^f$. It can be stated that the fundamental problem in geometrical optics is that given initial coordinates $\vec{r}^i$ and final coordinates $\vec{r}^f$, how does one design a system such that initial and final ray quantities have specific predetermined properties. Since the rays must in general satisfy Fermat's principle, the relationship between initial and final states cannot be in general arbitrary. A common method is to use Hamilton's characteristic function methods[9,10]. It can be shown that for any arbitrary optical system there exists a single function of 6 variables that completely describes the relation between initial and final ray quantities. For example, there are general relations

$$
\begin{aligned}
n^i\vec{s}^i &= -\frac{\partial V}{\partial r^i} \\
n^f\vec{s}^f &= \frac{\partial V}{\partial r^f}
\end{aligned}
\tag{1}
$$

where $V(\vec{r}^i,\vec{r}^f)$ is the point-characteristic function, and n^i and n^f are the refractive indices of the media before and after the optical system. Even though these characteristic functions are important and frequently applied there are certain problems associated with them. For example, no one kind of characteristic function is applicable to all forms of optical systems. As Dragt [5] points out, a point-characteristic function is not applicable to an imaging system and an angle characteristic is not applicable to a telescopic system. It is also known that it is difficult to find explicitly the characteristic function for a compound system even when the characteristic functions of it's components are known.

The Lie approach to the dynamics of any Hamiltonian system depends on the symplectic map. That is, if changes are made in the initial coordinates, there will be corresponding changes in the final coordinates, the relationship between the two being given by a mapping. In the case of optical imaging, we recall that the optical path length form p^i to p^f is given by the integral:

$$\int n ds = \int n(x,y,z)\sqrt{1+(x')^2+(y')^2}\,dz \tag{2}$$

Where the symbols have the usual meanings.

From Fermat's principle, we know that the ray path satisfies the Euler-Lagrange equation

$$\frac{d}{dz}\left(\frac{\partial L}{\partial x'}\right)-\frac{\partial L}{\partial x}=0$$

$$\frac{d}{dz}\left(\frac{\partial L}{\partial y'}\right)-\frac{\partial L}{\partial y}=0 \tag{3}$$

where, $L=\int n(x,y,z)\sqrt{1+(x')^2+(y')^2}\,dz$ is the optical Lagrangian. By introducing as done previously the two conjugate momenta

$$p_x=\frac{\partial L}{\partial x'} \qquad p_y=\frac{\partial L}{\partial y'} \tag{4}$$

we can derive the corresponding Hamiltonian function

$$H=-\left[n^2(\vec{r})-p_x^2-p_y^2\right]^{\frac{1}{2}} \tag{5}$$

Now, define a four-component vector $\vec{w}$

$$\vec{w} = (w_1, w_2, w_3, w_4) = (q_x, q_y, p_x, p_y) \tag{6}$$

and $\vec{w}^i$ and $\vec{w}^f$ the initial and final co-ordinates. The relationship between the initial and final co-ordinates can be described as a mapping ***M***. This is also called an optical transfer map. Fermat's principle implies that the light ray trajectory is given by the Hamiltonian and, ***M*** is a symplectic map. If we let M be a Jacobian matrix defined by

$$M_{a,b} = \frac{\partial w_a^f}{\partial w_b^i} \tag{7}$$

and J be a 4 x 4 matrix

$$J = \begin{bmatrix} 0 & I \\ -I & 0 \end{bmatrix} \tag{8}$$

where I is the 2 x 2 identity matrix, then it can be shown [9; see Appendix C] that,

$$M^T JM = J \tag{9}$$

where M^T is the transpose of M. This is the condition that **M** is a symplectic matrix. In optics the above equation leads to the well-known lens equation and is itself called the Lens Equation and **M** is the optical symplectic map. A symplectic map is a canonical transformation and vice-versa.

The above can be generalized easily to the set of 2n phase space coordinates. If ζ^i and ζ^f are initial and final coordinates of a trajectory governed by a Hamiltonian

$$\zeta^f = M\zeta^i \tag{10}$$

and

$$\zeta_a^f = k_a + \sum_b R_{ab}\zeta_b^i + \sum_{b,c} T_{abc}\zeta_b^i\zeta_c^b + \ldots \tag{11}$$

$$M_{a,b} = \frac{\partial \zeta_a^f}{\partial \zeta_b^i} \qquad (12)$$

and the J matrix is 2n x 2n matrix.

The symplectic map as defined above can be uniquely defined in terms of a Lie transformation operator. Here we introduce (without proof) Lie algebraic tools: The reader is referred to the literature for details. If f and g are any two functions defined on the phase space generalized co-ordinates and momenta, then we can define the Poisson bracket as:

$$\{f, g\} = \sum_i \left(\frac{\partial f}{\partial q_i} \frac{\partial g}{\partial p_i} - \frac{\partial f}{\partial p_i} \frac{\partial g}{\partial q_i} \right) \qquad (13)$$

We introduce a Lie operator that is denoted by $:f:$ which acts on general functions g and is defined in terms of Poisson brackets by

$$:f:g = \{f, g\} \qquad (14)$$

and powers of $:f:$ are defined by taking repeated Poisson brackets

$$\begin{aligned} :f:^0 g &= g \\ :f:^2 g &= \{f, \{f, g\}\} \\ :f:^3 g &= \{f, \{f, \{f, g\}\}\} \end{aligned} \qquad (15)$$

Next, we deal with power series of $:f:$. Specifically, we define the power series of $\exp(:f:)$. This is called the Lie transformation associated with $:f:$ and is defined by

$$\exp(:f:) = \sum_{n=0}^{\infty} \frac{:f:^n}{n!} \qquad (16)$$

with the rule

$$\exp(:f:)g = g + \{f,g\} + \frac{1}{2}\{f,\{f,g\}\} + \frac{1}{3!}\{f,\{f,\{f,g\}\}\} + \ldots \tag{17}$$

A fundamental result known as the Factorization theorem has been derived by Dragt and Finn [11] This theorem basically states that if **M** be a symplectic map between an initial phase space variable, and a final phase space variable, then the mapping can be uniquely written as a product.

$$M = \exp(:f_2:)\exp(:f_3:)\exp(:f_4:)\ldots. \tag{18}$$

Where f_n is a polynomial of degree n in the components of $\vec{w}$. This solution completely satisfies the Lens equation. The above equation is valid when the mapping maps the origin of phase space into itself. The physical meaning is that the effect of any collection of optical elements (lenses, mirrors, prisms, etc.) can be characterized by a set of homogenous polynomials. The polynomial f_2 represents paraxial optics, $f_3, f_4,$ and other higher order functions represent departures from paraxial optics and characterize the non-linear effects. For example, f_4 represents third order aberrations, f_6, fifth order aberrations, etc. when we consider a system with axial rotation and reflection symmetries.

The crucial point to note is that even if we truncate the product at a finite number of steps, the product still represents a symplectic map. As stated by Dragt "from Lie algebraic perspective the fundamental problem in geometrical optics is to study what polynomials correspond to various optical elements, to study what polynomials result from concatenating various optical elements and to study what polynomials correspond to various desired optical properties"[5].

Some examples

It is easy to show that if we retain only the second order polynomials f_2, then all the results of the Gaussian approximation can be reproduced by a suitable choice of f_2.

If f_2 is chosen to be:

$$f_2 = \frac{-\ell}{2n}\left(\vec{p}^{\,i}\right)^2 \tag{19}$$

Then, with some algebraic manipulation

$$\begin{aligned}\vec{q}^{\,f} &= M\vec{q}^{\,i} = \exp\left(-\frac{\ell}{2n} : \vec{p}^{\,2}\right)\vec{q}^{\,i} \\ &= \vec{q}^{\,i} + \frac{\ell}{n}\vec{p}^{\,i}\end{aligned} \tag{20}$$

$$\vec{p}^{\,f} = \mathbf{M}\vec{p}^{\,f} = \vec{p}^{\,i}$$

This is nothing but a transit through a medium of uniform refractive index n, through a distance ℓ.

Next, consider the choice of

$$f_2 = \frac{n_2 - n_1}{2r}\vec{q}^{\,i2} \tag{21}$$

Working out the required Poisson brackets will show

$$\vec{q}^{\,f} = \mathbf{M}\,\vec{q}^{\,i} = \exp(:f_2:)q^i = \vec{q}^{\,i} \tag{22}$$

and

$$\vec{p}^{\,f} = M\;\vec{p}^{\,i} = \exp(:f_2:)p^i = \vec{p}^{\,i} + \frac{n_2 - n_1}{2r}\vec{q}^{\,i} \tag{23}$$

This is obviously nothing but the result predicted for the case of light ray traveling in a medium on index n_1, incident on a spherical surface of radius of curvature r, separating a medium of index n_2 Similarly, it is easy to show that if

$$f_2 = -\sigma\left(\vec{p}^{\,i}\cdot\vec{q}^{\,i}\right) \tag{24}$$

where σ is some parameter. Then,

$$\vec{p}^{f} = \left(e^{\sigma}\right)\vec{q}^{i}$$

$$p^{f} = \left(e^{-\sigma}\right)\vec{p}^{i} \tag{25}$$

These can be recognized as equations for a system that is both imaging and telescopic with magnification exp(σ).

In general, if we assume that the optical system is axially symmetric with some axis and is also symmetrical with respect to reflection through some plane containing the axis of symmetry, then, it can be shown from symmetry considerations that only even indices of the polynomial f_n exist. In other words only polynomials containing the terms of $\vec{p}^2$, $\vec{p}\cdot\vec{q}$ and $\vec{q}^2$ will exist, with the odd power polynomials being equal to zero. In the general case of a completely asymmetric system, the other odd-powers will exist. The polynomials f_4, f_6, etc. will correspond to 3rd, 5th, etc. order aberrations.

Third order Aberrations

Let us consider just the third-order aberration. Then we can write the most general homogenous polynomial as:

$$f_4 = A(\vec{p})^2 + B\vec{p}^2(\vec{p}\cdot\vec{q}) + C(\vec{p}\cdot\vec{q})^2 + Dp^2q^2 + E(\vec{p}\cdot\vec{q})\vec{q}^2 + F\left(\vec{q}^2\right)^2 \tag{26}$$

The quantities A-F are arbitrary co-efficients whose values depend on the particular optical system. The explicit values can be calculated using the Baker-Campbell-Haausdorff theorem, a useful computational theorem used in Lie algebra. Essentially the theorem states that if A and B are 2 exponential matrices, then the product of the exponentials e^A e^B is e^C. C can be expressed as a series containing A,B and multiple commutations of A and B. Further discussion is beyond the scope of the chapter. If we compare the above expression with the general Taylor series representation of the wavefront (up to third order), then we can immediately note that the quantities A-F are the various Seidel aberrations. These aberrations are enumerated and explicit expressions in terms of the generalized coordinates $(\alpha = 1,2)$ are given below.

1. Spherical Aberration

$$\begin{aligned} f &= A(p^2)^2, \\ q_\alpha^f &= q_\alpha^i - 4Ap_\alpha^i p^{i2}, \\ p_\alpha^f &= p_\alpha^i \end{aligned} \tag{27}$$

2. Coma

$$\begin{aligned} f &= Bp^2(p\cdot q), \\ q_\alpha^f &= q_\alpha^i - B\left[2p_\alpha^i(p^i\cdot q^i)+p^{i2}q_\alpha^i\right] \\ &+\frac{B^2}{2!}\left[4p^{i2}(p^i\cdot q^i)p_\alpha^i - p^{i4}q_\alpha^i\right] \\ p_\alpha^f &= p_\alpha^i + Bp^{i2}p_\alpha^i + \frac{B^2}{2!}3p_\alpha{}^{i4}p_\alpha^i \end{aligned} \tag{28}$$

Terms of order higher than B^2 have been ignored.

3. Astigmatism

$$\begin{aligned} f &= C(p\cdot q)^2, \\ q_\alpha^f &= q_\alpha^i - 2C(p^i\cdot q^i)q_\alpha^i + 2C^2(p^i\cdot q^i)^2 q_\alpha^i, \\ p_\alpha^f &= p_\alpha^i + 2C(p^i\cdot q^i)p_\alpha^i + 2C^2(p^i\cdot q^i)^2 p_\alpha^i \end{aligned} \tag{29}$$

We have neglected terms of the order of C^3 and higher.

4. Curvature of Field

$$f = Dp^2q^2,$$
$$q_\alpha^f = q_\alpha^i - 2Dq^{i2}p_\alpha^i - 2D^2\left[p^{i2}q^{i2}q_\alpha^i - 2q^{i2}(p^i \cdot q^i)p_\alpha^i\right] \tag{30}$$
$$p_\alpha^f = p_\alpha^i + 2Dp^{i2}q_\alpha^i + 2D^2p^{i2}\left[2(p^i \cdot q^i)q_\alpha^i - q^{i2}p_\alpha^i\right]$$

Here we have neglected terms of order D^3.

5. Distortion

$$f = E(p \cdot q)q^2,$$
$$q_\alpha^f = q_\alpha^i - Eq_\alpha^i q^{i2} + 3\frac{E^2}{2!}q^{i4}q_\alpha^i, \tag{31}$$
$$p_\alpha^f = p_\alpha^i + E\left[p_\alpha^i q^{i2} + 2q_\alpha^i(p^i \cdot q^i)\right]$$
$$+\frac{E^2}{2!}\left[4q^{i2}(q^i \cdot p^i)q_\alpha^i - q^{i4}p_\alpha^i\right]$$

Terms of order E^3 and higher have been neglected

6. This sixth term has no effect on the quality of the image. However, it affects the arrival direction α of the ray in the image plane.

$$f = F(q^2)^2,$$
$$q_\alpha^f = q_\alpha^i, \tag{32}$$
$$p_\alpha^f = p_\alpha^i + 4Fq_\alpha^i q^{i2}.$$

Some additional examples

Example 1: Consider a lens with planar entrance and exit faces (figure 8.1). Let the refractive index of the surrounding medium be n_1 and of the lens be n_2. The z-axis is along the optical axis and x and y axis along the face of the lens. Let the shape of the curved interface be given by:

$$z = z_2(q) = -\beta_2\left(q_1^2 + q_2^2\right) + \beta_4\left(q_1^2 + q_2^2\right)^2 + \ldots\ldots \tag{33}$$

or in Cartesian coordinates

$$z = -\beta_2\left(x^2 + y^2\right) + \beta_4\left(x^2 + y^2\right)^2 + \beta_6\left(x^2 + y^2\right)^3 + \ldots \tag{34}$$

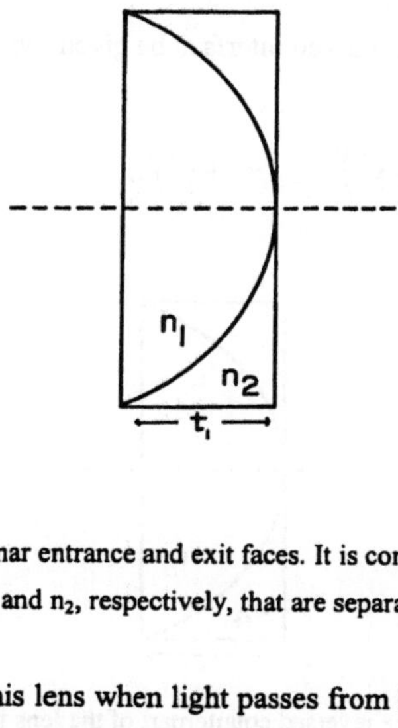

Figure 8.1: Lens with planar entrance and exit faces. It is composed of two media having indices of refraction n_1 and n_2, respectively, that are separated by a curved interface.

The symplectic map for this lens when light passes from left to right is given through fourth order by the following expression:

$$M = \exp\left(\frac{-t_1}{2n_1} : \vec{p}^2 :\right) \exp\left(\frac{-t_1}{8n_1^{\,3}} : \left(\vec{p}^2\right)^2 :\right) \exp\left(\beta_2\left(n_2 - n_1\right) : \vec{q}^2\right) \exp(: f_4 :) \tag{35}$$

where f_4 has the general expansion given above. With some algebraic manipulation it is easy to show that the coefficients are

$$A = 0, B = 0, C = 0$$

$$D = \beta_2 \frac{n_2 - n_1}{2n_1 n_2}$$

$$E = -2\beta_2^{\,2} \frac{n_1 - n_2}{n_1} \tag{36}$$

$$F = \frac{1}{n_1}\beta_2^{\,3}(n_1 - n_2)\left(n_1\left[\frac{\beta_4}{\beta_2^{\,3}} + 2\right] - 2n_2\right)$$

Now, consider the reversed counterpart of the lens discussed above (see Figure 8.2). Let t_2 be the thickness of the lens and n_3 be the refractive index of the material of the lens . Let the shape of the curved interface be given by:

$$z = \gamma_2(x^2 + y^2) - \gamma_4(x^2 + y^2)^2 - \gamma_6(x^2 + y^2)^3 \ldots \tag{37}$$

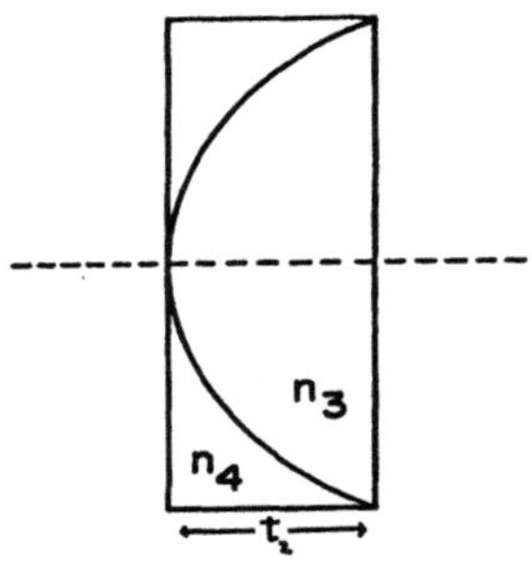

Figure 8.2: The reversed counterpart of the lens in figure 8.1

It is obvious that passage of light rays from left to right is related to passage of light rays from right to left in the reversed counterpart. Additionally, the passage of light rays from right to left is the reverse process of transit from left to right through the same lens. We can write the symplectic map for the reverse counterpart as:

$$M = \exp(:g_4:)\exp(\gamma_2(n_4 - n_3):\vec{q}^2:)\exp\left(\frac{-t_2}{2n_3}:\vec{p}^2:\right)\exp\left(\frac{-t_2}{8n_3^{\,3}}:(\vec{p}^2)^2:\right) \tag{38}$$

Where g_4 has an expansion similar to f_4 with the coefficients being

$$
\begin{aligned}
&A = 0, B = 0, C = 0 \\
&D = \gamma_2 \frac{n_4 - n_3}{2n_3 n_4} \\
&E = -2\gamma_2{}^2 \frac{n_3 - n_4}{n_3} \\
&F = \frac{1}{n_3}\gamma_2{}^3 (n_3 - n_4)\left(n_3 \left[\frac{\gamma_4}{\gamma_2{}^3} + 2 \right] - 2n_4 \right)
\end{aligned}
\tag{39}
$$

These results for the coefficients are very similar, with n_1 and n_2 being replaced by n_3 and n_4 and β_2 and β_4 replaced by γ_2 and γ_4. Additionally, the orders of the factors in the mapping have been reversed and the sign of the coefficient E has been changed. It is easy to verify that

$$
g_4(\vec{q}, \vec{p}) = f_4(\vec{q}, -\vec{p}) \tag{40}
$$

We can now create a lens with 2 curved surfaces (figure 8.3). If we join together the two lens segments shown above, with setting $n_4 = n_1 = 1$ and $n_3 = n_2 = n$, and $t = t_1 + t_2$. The symplectic map is then simply the product of the two mappings given above.

$$
\begin{aligned}
M = &\exp(:g_4:)\exp(\beta_2(1-n):\vec{q}^2:)\exp\left(\frac{-t}{2n}:(\vec{p}^2)^2:\right)\exp\left(\frac{-t}{8n^3}:\vec{p}^2:\right) \\
&\bullet \exp(r_2(1-n):\vec{q}^2:)\exp(:f_4:)
\end{aligned}
\tag{41}
$$

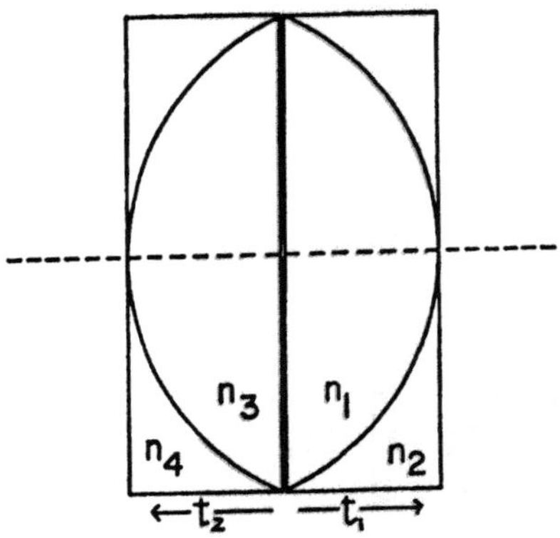

Figure 8.3: A compound lens formed by joining together the lenses of Figures 8.1 and 8.2.

In the Gaussian approximation, leaving out higher order terms characterizing aberrations, we get

$$M = \exp\left(\beta_2(1-n):\vec{q}^{\,2}:\right)\exp\left(\frac{-t}{2n}:\vec{p}^{\,2}:\right)\exp\left(\gamma_2(1-n):\vec{q}^{\,2}:\right) \tag{42}$$

The effect of this map is:

$$\begin{pmatrix}\vec{q}^{\,f}\\ \vec{p}^{\,f}\end{pmatrix} = M\begin{pmatrix}\vec{q}^{\,i}\\ \vec{p}^{\,i}\end{pmatrix} = \varsigma\begin{pmatrix}\vec{q}^{\,i}\\ \vec{p}^{\,i}\end{pmatrix} \tag{43}$$

where ς is simply the ray-transfer matrix[12], and is given by

$$\varsigma = \begin{bmatrix}A & B\\ C & D\end{bmatrix} \tag{44}$$

The power of the optical system is given by $P=-C$. Evaluating, the matrix elements explicitly, we get

$$\varsigma = \begin{bmatrix} A & B \\ C & D \end{bmatrix}$$

$$= \begin{bmatrix} 1+\dfrac{(n-1)t}{nR_1} & \dfrac{t}{n} \\ (1-n)\left[\left(\dfrac{1}{R_1}-\dfrac{1}{R_2}\right)+\dfrac{(n-1)t}{nR_1R_2}\right] & 1-\dfrac{(n-1)t}{nR_2} \end{bmatrix} \tag{45}$$

The power is given by

$P = -C = 1/f$ where f is the focal length

$$= (n-1)\left[\left(\frac{1}{R_1}-\frac{1}{R_2}\right)+\frac{(n-1)t}{nR_1R_2}\right] \tag{46}$$

Here, we have made the identification $\gamma_2 = -\frac{1}{2R_1}$, and $\beta_2 = \frac{1}{R_2}$, where and R_2 are radii of curvature of the 2 surfaces. The above expression for power is nothing but the equivalent power of a thick lens, with surfaces having radii of curvature R_1 and R_2 and thickness t. For a thin lens, t=0 and we get the usual lensmaker's equation.

Example 2: Correction of Spherical Aberration

Following Rangarajan and Sachidanand [13] we show how spherical aberration can be corrected using Lie algebraic techniques. As before, consider light propagating through a lens of index n, in a medium of index n_2 =1.00, having surfaces parameterized by the equation (figure 8.4)

$$z^s = \beta_2(x^2+y^2)+\beta_4(x^2+y^2)^2+\beta_1(x^2+y^2)^3+\beta_8(x^2+y^2)^4+\ldots.. \tag{47}$$

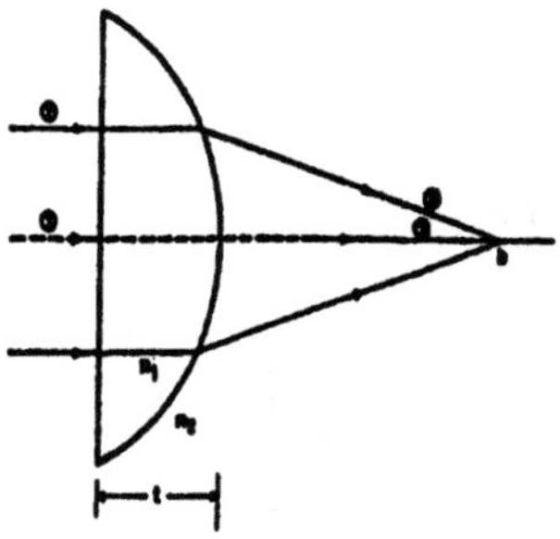

Figure 8.4: A simple imaging system

We can analyze this situation as having three distinct parts:

1. passage through a distance t in a medium of index n, the lens
2. surfaces of lens separating media of refractive indices n and 1.
3. transmit through a distance l, the focal length in air.

The transfer maps for 1 and 3 are given by expressions of the form for transit through, a slab of thickness l and index n.

$$M = \exp\left\{ l : \left(n^2 - \vec{p}^{\,2}\right)^{\frac{1}{2}} : \right\} \tag{48}$$

which is nothing but

$$M = \exp\left(-\frac{l}{2n} : \vec{p}^{\,2} : \right) \exp\left(-\frac{l}{8n^3} : \left(\vec{p}^{\,2}\right)^2 : \right) \exp\left(-\frac{l}{16n^5} : \left(\vec{p}^{\,2}\right)^3 : \right)$$
$$\cdot \exp\left(-\frac{5l}{128} : \left(\vec{p}^{\,2}\right)^4 : \right) \ldots \tag{49}$$

The map for (2) is given by:

$$\exp(:f_2:)\exp(:f_4:)\exp(:f_6:)\exp(:f_8:)\ldots \tag{50}$$

where

$$f_2 = \beta_2 (n_1 - n_2) \vec{q}^2 \tag{51}$$

$$f_4 = -\beta_2^3 \left(\frac{n_1 - n_2}{n_1} \right) \left\{ n_1 \left[2 - \frac{\beta_4}{\beta_2} - 2n_2 \right] \right\} (\vec{q}^2)^2 + 2\beta_2 \left(\frac{n_1 - n_2}{n_1} \right) (\vec{q}^2)(\vec{p} \cdot \vec{q})$$
$$+ \frac{\beta_2 (n_1 - n_2)}{2n_1 n_2} \vec{q}^2 \vec{p}^2$$

Coming back to our problem of transit through a lens, the combined transfer map is simply the product of the maps for the individual components.

$$M = M_1 M_2 M_3 \tag{52}$$

or,

$$M = \exp(:g_2:)\exp(:g_4:)\exp(:g_6:)\exp(:g_8:).. \tag{53}$$

where g_2, g_4 are of the form:

$$\begin{aligned} g_2 &= A_2 \vec{p}^2 + \beta_2 \vec{p} \cdot \vec{q} + C_2 \vec{q}^2 \\ g_4 &= A_4 (\vec{p}^2)^2 + \beta_4 (\vec{p} \cdot \vec{q}) \vec{p}^2 + \ldots \\ g_6 &= A_6 (\vec{p}^2)^3 + \beta_6 (\vec{p} \cdot \vec{q})(\vec{p}^2)^2 + \ldots \\ g_8 &= A_8 (\vec{p}^2)^4 + \beta_8 (\vec{p} \cdot \vec{q})(\vec{p}^2)^3 + \ldots \end{aligned} \tag{54}$$

The expressions for A_2, B_2 , C_2, A_4, etc. have been left out for brevity. These coefficients would depend on l and n. The co-efficients in g_2 would depend on β_2, coefficients in g_4 on β_2 and β_4, and so on.

We know that g_2 represents paraxial optics. This implies that a parallel beam of light incident on lens will come to a focus at a distance l along the z-axis

$$\begin{pmatrix} 0 \\ \vec{p} \end{pmatrix} = e^{:g_2:} \begin{pmatrix} \vec{q} \\ 0 \end{pmatrix} \tag{55}$$

This results in

$$\beta_2 = \frac{1}{2l(1-n)} \tag{56}$$

since l and n in g_2 are already specified. The other coefficients β_4, β_6, and β_8 can be fixed by imposing conditions such that M is free of spherical aberration up to order 7. Spherical aberration is defined conventionally as a displacement from the paraxial focus by an amount dependent only on the arrival angle $(\vec{p})$.

It is also seen from our previous discussion only terms $(\vec{p}^2)^2$, $(\vec{p}^2)^3$ and $(\vec{p}^2)^4$ in g_4, g_6, and g_8 give rise to spherical aberration of orders 3,5, and 7. To see that is true, consider g_4.

$$\begin{aligned} \vec{q}^f &= \exp\left(:A_4(\vec{p}^2)^2:\right)\vec{q}^i \\ &= \vec{q}^i + A_4\left\{(\vec{p}^2)^2, q^i\right\} \\ &= \vec{q}^i - 4A_4(\vec{p}^2)\vec{p} \end{aligned} \tag{57}$$

and

$$\Delta\vec{q} = \vec{q}^f - \vec{q}^i = -4A_4(\vec{p}^2)\vec{p} \tag{58}$$

The actual arrival point $\vec{q}^{\,f}$, arrival point's displacement from Gaussian arrival point $\vec{q}$ depends only on the arrival direction $\vec{p}$, and hence is spherical aberration.

For spherical aberration to be zero, $\Delta\vec{q} = 0$. This can be achieved by setting $A_4 = A_6 = A_8 = 0$ (up to 7th order) and solve the equations for β_4, β_6, β_8 in terms of n and l. After algebraic manipulations, we can show that:

$$\beta_4 = \frac{(n+l)}{8l^3(n-l)^2}$$
$$\beta_6 = \frac{-(n+l)^2}{16l^5(n-l)^3} \tag{59}$$
$$\beta_8 = \frac{5(n+l)^3}{128l(n-l)^4}$$

and the resultant equation of the lens surface that would correct spherical aberration up to 7th order is

$$z = -\frac{1}{21(n-1)}y^2 + \frac{(n+1)}{81^3(n-1)^2}y^4 - \frac{(n+1)^3}{161^5(n-1)^3}y^6 + \frac{5(n+1)^3}{1281^7(n-1)^4}y^8 \tag{60}$$

Solution using Fermat's principle:

Consider figure 8.4. Fermat's principle requires that rays 1 and 2 should have the same ray path. This implies in the y-z plane:

$$n(t+z) + \sqrt{(1-z)^2 + y^2} = nt + l \tag{61}$$

where l is the focal length (secondary) of the lens, n is the refractive index and t is the thickness.

Solving the above equation for Z and using the physically appropriate solution namely that as y increases, there is a decrease in Z

$$Z = \frac{t}{n+1}\left[1 - \sqrt{1 + \frac{(n+1)y^2}{(n-1)l^2}}\right] \tag{62}$$

This is nothing but the equation of a hyperbola. Expanding the square root

$$Z = \frac{-y^2}{2l^2(n-1)} + \frac{(n+1)y^4}{8l(n-1)^2} - \frac{(n+1)^2 y^6}{16l^5(n-1)^3} + \frac{5(n+1)^3 y^8}{128l^7(n-1)^4} - \ldots\ldots \tag{63}$$

Comparing equations (63) and (61) we note they are identical. These equations give the exact shape of the lens that is free of on-axis spherical aberration.

Example 3: A general example: extension to gradient-index media

As before consider propagation of a light ray in a medium that has a constant refractive index n_1, hits an interface given by a surface and another medium of index n_2.

That is

$$n(z) = n_1 + (n_2 - n_1)\theta\left[z - \beta_2\left(q_1^2 + q_2^2\right) - \beta_4\left(q_1^2 + q_2^2\right)^2 \ldots\right] \tag{64}$$

Where θ is a Heaviside step function. Therefore, the Hamiltonian is completely known. The evolution of $\zeta^i = (q_i, p_i) \to \zeta^f = (q_f, p_f)$ is governed by the Hamiltonian, the map $\zeta^i = M\zeta^f$ is given by

$$\frac{dM}{dz} = M : -H(\zeta^i, z): \tag{65}$$

The unique solution to this equation for a given Hamiltonian can be written as.

$$M = P\left\{\exp\left[-:\int_{z^i}^{z} H(\zeta', z')dz' :\right]\right\} \tag{66}$$

where P represents z ordering of the exponential. For a gradient index medium that is axially symmetric, we just need to write

$$n(z) = n_o(z) + n_2(z)(q_1^2 + q_2^2) + n_4(z)(q_1^2 + q_2^2) + \ldots \quad (67)$$

which completely determines the Hamiltonian.

Because the mapping from initial to final conditions in any optical system is a symplectic mapping, they can be analyzed using techniques of Lie algebra. This technique utilizing the optical Lagrangian is very powerful and have found applications in many fields such as accelerator design [7,14] visual optics[15], and electron microscopy [16] This technique has been extended to the case of optical systems with small misalignment and misplacement errors also. A computer code MARYLIE a program for charged particle beam transport based on lie methods has been developed by Prof. A.J.Dragt and his group at the Univ. Of Maryland, College Park. MD.

Without going into details, it can also be shown that various features of the Lie algebra can be related to the associated Lie group Sp(2) or sp(2) and specific computational tools associated with the three-dimensional rotation group[SU(2)], can be applied to problems concerning Sp(2) and sp(2). The Lie algebra sp(2) can be used to classify departures from paraxial optics. This, however, is beyond the scope of this chapter.

How does the Lie algebraic method compare with other, more common, techniques? There are several computer ray-tracing programs that trace a large number of rays by employing Snell's law at various interfaces in the system under consideration and by searching the lens parameter space for optimization of the various features of the resulting spot diagram. However, as pointed out by Dragt [5,7], these search codes tend to run rather blindly. Lie algebraic techniques may provide additional information regarding the sources of aberrations and the extent to which these may be corrected. We would like to end this comparison of Lie techniques vis-a-vis other methods by quoting Hawkes[12], who in the context of charged-particle optics wrote,

"This said, it is in no sense my intention to provoke a polemic between the proponents of Lie optical methods and some (perhaps already fictitious), old guard of eikonal lovers: Having myself used characteristic functions for more than a quarter century, I am delighted to find that familiarity with the Lie approach has enabled me to solve at least one long-standing problem. I am convinced that these different approached must henceforward be regarded as complementary, some problems being easier to treat by one rather than the other and in practice, better software often being available one than for the other. Moreover, for complex systems, it is certainly wise to use both methods in order to check the results. It is therefore highly desirable that optical designers should be familiar with both and that the two approaches should be described fully in future texts on aberrations".

A full-scale comparison between Lie methods and conventional methods for general light-optical systems is not yet available, to the best of our knowledge.

References

1. M. Kando and Y. Kateuchi, *Matrix method for nonlinear transformations and its application to an optical system*. J. Opt. Soc. Am. A, **13**, 71, (1996)
2. V. Lakshminarayanan and S. Varadharajan, *Expressions for aberration coefficients using non-linear transformations*. Optom. Vis. Sci., **74**, 676 (1997)
3. V. Lakshminarayanan and S. Varadharajan, *Aberration coefficients of general spherocylindrical surfaces*. Optom. Vis. Sci., **77**, 156, (2000)
4. J.B. Almeida, *General methods for the determination of matrix coefficients for high order optical system modeling*, J. Opt. Soc. Am.A, **16**, 596, (1999)
5. A.H. Dragt, *Lie algebraic theory of geometrical optics and optical aberrations*, J. Opt. Soc. Am, **72**, 372, (1982)
6. A.J. Dragt, E. Forest, K.B. Wolf, *Foundations of a Lie algebraic theory of geometric optics* in *Lie Methods in Optics*, Vol. 250 of lecture notes in physics, ed. J Sanchez-Mondragon and K.B. Wolf., Springer Verlag, Heidelberg, (1986). Pages 105-107. This book contains extensive overview of Lie group theory and applications
7. A.J. Dragt, *Elementary and advanced Lie algebraic methods with application to accelerator design, electron microscopes and light optics*. Nucl. Instrum. Meth. Phys. Res. A, **258**, 339, (1987)
8. V. Guillemin, S. Sternberg, *Symplectic techniques in Physics*, Cambridge Univ. Press, Cambridge, UK, (1984)
9. O.Staroudis, *The optics of rays, wavefronts and caustes*, Academic Press, NY, (1972)
10. R.K. Luneberg, *Mathematical theory of optics*, Univ. of California Press, Berkeley, (1964)
11. A.J.Dragt and J.M. Finn, *Lie series and invariant functions for analytic symplectic maps*, J. Math. Phys., **17**, 2215, (1976)
12. A. Gerrard and M. Burch, *Introduction to matrix methods in optics*, Dover, NY, (1994)
13. G. Rangarajan and M. Sachidanand, *Spherical aberration and its correction using lie algebra techniques*, Pramana-Ind. J. Phys., **49**, 635, (1997)
14. E. Forest, *Lie algebraic methods for charged particle beams and light optics*, Ph.D. dissertation, Univ. Maryland, College Park, MD. (1984)

15. V. Lakshminarayanan, R. Sridhar, R. Jagannathan, *Lie algebraic treatment of dioptric power and optical aberrations*. J. Opt. Soc. Am. A, **15**, 2497, (1998)

16. S.A. Khan, and R. Jagannathan, *Quantum mechanics of charged-particle beam transport through magnetic lenses*, Phys. Rev.,E, **51**, 2510, (1995)

17. P.W. Hawkes, *Lie methods in optics: an assessment, in Lie methods in optics II*, K.B. Wolf, ed., Vol. 352 of Springer Lecture notes in Physics, Springer-Verlag, Heidelberg, (1989). Pages 1-17.

9. AN INTRODUCTION TO DYNAMIC PROGRAMMING AND APPLICATIONS TO OPTICS

Introduction

The use of optimization theory to describe light propagation phenomena through optical media with different optical properties and geometries is an area of great theoretical interest. This methodology concerns optimized design of arbitrary optical media for which an *a priori* characterization of significant parameters would be required. The problem of characterizing the output of a system from partial input data is defined as a control theoretical problem. The mathematical theory of adaptive control processes was first introduced by Bellman [1] and has been applied to wide variety of problems in Engineering, Physics, Biology, Economics and even Medicine [1-5]. The problem is usually stated as the minimization of a functional equation. An approach to solving this problem is a technique called dynamic programming which will be introduced in this chapter.

Dynamic Programming

Generally, speaking, dynamic programming is a procedure of optimization that can be applicable to problems that require a sequence of inter-related "decisions". Then, each decision transforms the current situation into a new one. A sequence of decisions, which is going to yield a sequence of situations maximizes/minimizes some measurable value. It is worth remembering that all physical processes are sequential in nature.

Let us consider the function

$$R(x_1, x_2, x_3 \ldots x_n) = g(x_1) + g(x_2) + \ldots + g(x_n) \tag{1}$$

we need to maximize/minimize this function over the region $x_i \geq 0$, $\sum_{i=1}^{n} x_i = X$. We are assuming that x is any positive value and n is any integer value. We also assume that the given function is time independent and, hence, all the decisions made within a region $x_i \geq 0$ will be made one at time. That is, the control process is a process where influences, i.e., "controls" may be exerted upon a system at various times. Ordinarily this means that decisions have been made as to the nature of these actions. We are going to use a dynamic allocation process: The maximum of the given function $R(x_1, \ldots x_n)$ over the designated region is dependent upon both x and n. Let's make this dependence more specific by introducing the sequence of functions $\{f_n(x)\}$ where n=1,2,...n and

$$f_n(x) = \max_{\{x_i\}} R(x_1, \ldots\ldots x_n) \tag{2}$$

The function $f_n(x)$ represents the optimal return from an allocation of the quantity of resources x to n activities. For the sequence $\{f_n(x)\}$,

$$f_n(0) = 0 \quad n=1,2,\ldots \tag{3}$$

which provides that for $V_i \in [1, n], g_i(o) = 0$
and hence

$$f_1(x) = g_1(x) \text{ for } x \geq 0 \tag{4}$$

Now we need to obtain a recurrence relation between $f_n(x)$ and $f_{n-1}(x)$. Let x_n be the allocation for the nth activity, where $0 \le x_n \le x$. Regardless of the precise value of x_n, the remaining quantity of sources $(x - x_n)$ will be used to obtain the maximum return from the remaining $(n-1)$ activities. Because the optimal return for $(n-1)$ activities starting with quantity $(x - x_n)$ is by definition $f_{n-1}(x - x_n)$, the initial allocation of x_n to the nth activity will result in a total return $g_n(x_n) + f_{n-1}(x - x_n)$ from the n-activity process.

An optimal choice of x_n is the one that is going to optimize this function. The basic functional equation is:

$$f_n(x) = \max_{0 \le x_n \le x} \left[g_n(x_n) + f_{n-1}(x - x_n)\right] \qquad (5)$$

for $n = 2,3,......x \ge 0$ and where $f_1(x) = g_1(x)$

Here we introduce the concept of policy: A policy is a rule for making a decision based upon the available knowledge of the state of system. A policy is equivalent to an algorithm acting upon state variables. An optimal policy is a policy which yields the best overall result. We now describe the "principle of optimality" which is used to convert the problem of determining an optimal policy into a conventional mathematical task of solving a functional equation. The intuitive principle is an "optimal policy has the property that whatever the initial state and initial decisions are, the remaining decisions must constitute an optimal policy with regard to the state resulting from the first decision".

Since

$$\max_{x_1+x_2+....x_n=x} = \max_{0\le x_n\le x}\left[\max_{x_1+x_2+....+x_{n-1}=x-x_n}\right]$$

$$f_n(x) = \max_{x_1+x_2+....x_n=x}\left(g_n(x_n)+g_{n-1}(x_{n-1})+....+g_1(x_1)\right) \tag{6}$$

$$= \max_{0\le x_n\le x}\left[\max_{x_1+x_2+....+x_{n-1}} g_n(x_n)+....+g(x_1))\right]$$

The sequence $\{f_n(x)\}$ can be obtained inductively if $f_1(x)$ is known. Let's now examine the way to obtain the sequence $\{f_n(x)\}$ computationally, if the recurrence relation is

$$f_n(x) = \max_{0\le x_n\le x}\left[g_n x(n) + f_{(n-1)}(x-x_n)\right]_{n-2,3,....x\ge 0} \tag{7}$$

with $f_1(x) = g_1(x)$

Consider the set of values $f_n(x)$ in the interval $[0, x_\circ]$. We are going to use the values at the finite grid $x = 0, \Delta, 2\Delta, \ldots, R\Delta = x_\circ$. Each element of the sequence $\{f_n(x)\}$ is going be evaluated at each(and only) these points. It is important to know the accuracy of interpolation because that's how values of $f_n(x)$ are going to be obtained for x values distinct from these grid points. If $k\Delta < x < (k+1)\Delta$, the simplest approximation is $f_n(x) = f_n(k\Delta)$, and the next simplest is one using a linear interpolation

$$f_n(x) = f_n(k\Delta) + (x-k\Delta)(f_n(k+1)\Delta - f_n(k\Delta))/\Delta \tag{8}$$

and when n=1,

$$f_1(x) = g_1(x) \tag{9}$$

The interval $[0,x]$ is replaced by a discrete set of values and we get a set of values $\{f_1(k\Delta)\}, k = 0,1,...R.$

Then for n=2,

$$f_2(x) = \max_{K=0,1,..R} [g_2(x) + f_1(x - x_2)] \tag{10}$$

$$f_2(x) = \max_{0 \le x_2 \le x} [g_2(x_2) + f_1(x - x_2)] \tag{11}$$

where $x = 0, \Delta, 2\Delta, ...R\Delta$

In order to begin a process of maximization, we need to evaluate and compare the two functions $g_2(0) + f_1(x)$ and $g_2(\Delta) + f_1(x - \Delta)$ The larger quantity is kept and the process is continued till we exhaust all possible K values. Eventually, this will yield $f_2(x)$ for a particular value of x. In the process of searching we will determine the value of $f_2(x)$ when $x = 0, \Delta, 2\Delta, ...R\Delta$ as well as values of x_2 at which the maximum of $f_2(x)$ is obtained according to the given formula. This process is continued for all n stages.

However, most of the time the quantity of interest is not the process itself, but some function of it and the problem is to optimize that function. More details can be found in references. It should be pointed out that dynamic programming is intimately connected to the calculus of variations.

Application to Optics

Consider light propagation in an inhomogeneous medium. The main goal is to determine the path of a ray of light through this media with the assumption that the path minimizes time, which is nothing but Fermat's principle. Thus, the path of the ray can be found by minimizing the optical path length.

Consider a medium of refractive index n(x,z) and light propagation is along the z-axis. Using Caratheodory's theorem "the optical path length of any ray is a minimum", out the infinite number of different pathways possible, the light ray is going to take a path S for which the physical situation is shown in Figure 9.1.

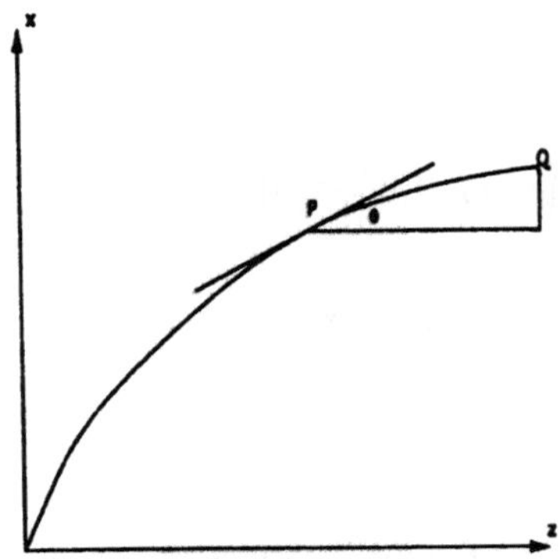

Figure 9.1: The geometry of the problem

$$S(x,z)+\min \int u(x(s),z(s))dS \tag{12}$$

Let the initial point be $(x_\circ, z_\circ)$ and for an arbitrary path S from $(x_\circ, z_\circ)$ to (x,z) we can write an intermediate stage $(x-\Delta x, z-\Delta z)$ to be the state resulting from the initial decision. For the arbitrary infinitesimal path ΔS from $(x-\sin\theta\Delta s, z-\cos\theta\Delta s)$ to (x,z), we can by applying reverse dynamic programming to show,

$$S(x,z)=\begin{matrix}\min\\ \theta\end{matrix}\begin{bmatrix}S(x-\sin\theta\Delta s, z-\cos\theta\Delta s)\\ +n(x-\sin\theta\Delta s, z-\cos\theta\Delta s)\Delta s+O(\Delta s)\end{bmatrix} \tag{13}$$

By applying Taylor's formula to the right side of the above equation,

$$S(x,z) = \min_{\theta} \begin{pmatrix} S(x,z) - \sin\theta\Delta s\left(\frac{\partial s}{\partial x}\right) - \cos\theta\Delta s\left(\frac{\partial s}{\partial x}\right) \\ + n(x,z)\Delta s + O(\Delta s) \end{pmatrix} = 0 \tag{14}$$

For $\Delta s \to 0$, and ignoring higher order terms,

$$\sin\theta\left(\frac{\partial s}{\partial x}\right) + \cos\theta\left(\frac{\partial s}{\partial z}\right) = n(x,z) \tag{15}$$

Let $S_x = \frac{\delta s}{\delta x}, S_z = \frac{\delta s}{\delta z}$ and $\tan\theta = \frac{S_z}{S_x}$

We have:

$$\sin\theta = \frac{S_x}{\sqrt{S_x^2 + S_z^2}}; \cos\theta = \frac{S_z}{\sqrt{S_x^2 + S_z^2}} \tag{16}$$

and

$$\begin{aligned} \sqrt{S_x^2 + S_z^2} &= \sqrt{n(x,z)\frac{\delta x}{\delta s}S_x + n(x,z)\frac{\delta z}{\delta s}S_z} \\ &= \sqrt{n(x,z)\left(\frac{\delta s}{\delta x}\cos\theta + \frac{\delta s}{\delta z}\sin\theta\right)} = n(x,z) \end{aligned} \tag{17}$$

Therefore we get

$$S_x^2 + S_z^2 = n^2(x,z) \tag{18}$$

which can readily be recognized as being the eikonal equation, the so-called fundamental equation of geometrical optics. The formalism of dynamic programming required the definition of a minimum pathway . The principle of optimality was applied to Fermat's principle. Thus the solution of the eikonal equation would give the optimized trajectory of a ray of light(Figure 9.2) To determine the optimized trajectory of a ray of light in the given media, we need to derive the function describing the dependence of z on x.

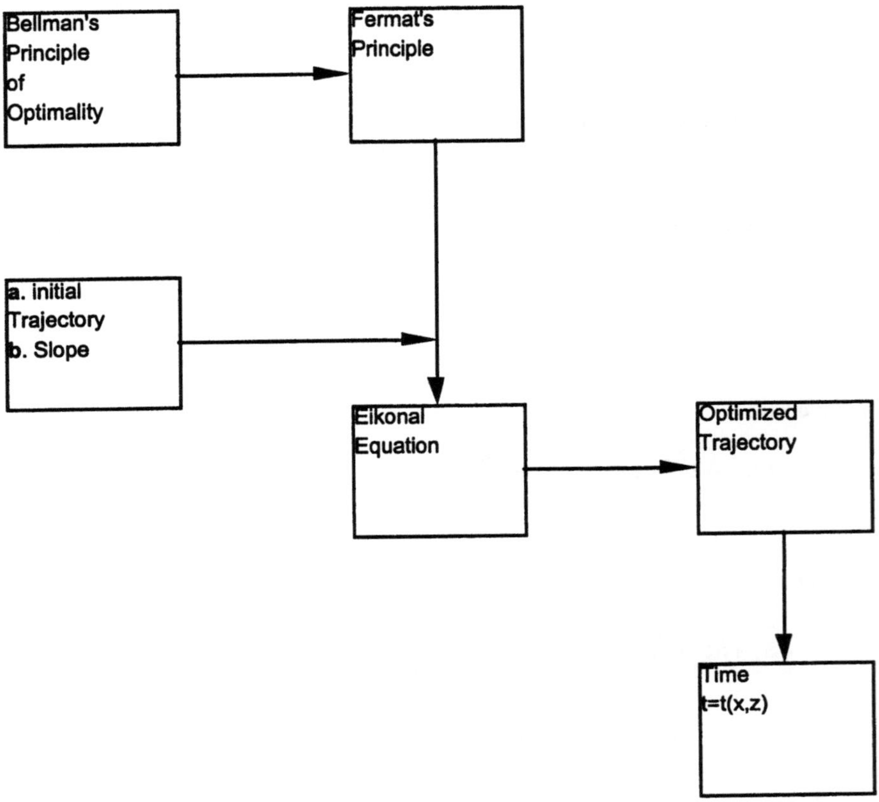

Figure 9.2: Schematic diagram showing the connection between dynamic programming and optimized trajectory (time minimization) for light propagating in an arbitrary optical medium.

Now,

$$dS^2 = dx^2 + dz^2 \qquad (19)$$
$$= \sqrt{1+(z'_x)^2}\,dx$$

Now consider the Lagrangian

$$L = n(x)\sqrt{1+(z'_x)^2}$$

(20)

$$\frac{\delta L}{\delta z'_x} = n(x)\frac{z'_x}{\sqrt{1+(z'_x)^2}} = n(x)\frac{\delta_z}{\delta_x\sqrt{1+(z'_x)^2}} = n(x)\frac{\delta z}{\delta S}$$

We define $\beta = n(x)\frac{\delta z}{\delta s} = (n(x)\cos\theta)$ which is simply the ray –invariant. This results in

$$dz = \frac{\beta dx}{\sqrt{n^2(x)-\beta^2}} \tag{21}$$

For a homogeneous medium, the index of refraction n is a constant, and the expression obtained shows that the path of a ray of light is going to be a straight line, as expected.

For an inhomogeneous medium, consider the simple case of $n(x) = x^2$

$$dz = \frac{dx}{\sqrt{\frac{x^4}{\beta^2}-1}} \tag{22}$$

integrating both sides, and making proper substitutions, this reduces to:

$$\int\frac{dx}{\sqrt{\frac{x}{\beta^2}-1}}=\sqrt{\beta}\int\frac{d(\frac{x}{\sqrt{\beta}})}{\sqrt{\left(\frac{x}{\sqrt{\beta}}\right)^4-1}}=\sqrt{B}\int\frac{dy}{\sqrt{y^4-1}}$$

$$=\sqrt{\beta}\int\frac{dy}{y^2\sqrt{1-\frac{1}{y^4}}}$$

As a first order approximation, we can write

$$Z\approx\sqrt{\beta}\int\frac{dy}{y^2\left(1-\frac{1}{2y^4}\right)} \tag{23}$$

$$z=-\sqrt{\beta}\sqrt[4]{2}\int\frac{dv}{1-v^4} \tag{24}$$

The integrand can be expressed in terms of partial fractions

$$\frac{1}{1-v^4}=\frac{a}{1-v}+\frac{b}{1+v}+\frac{cv+d}{1+v^2}$$

with

$$a=b=\frac{1}{4};c=0,d=\frac{1}{2}$$

and the final integral works out to be

$$z=\frac{-\sqrt{\beta}\sqrt[4]{2}}{4}\left[\ln\frac{\sqrt[4]{2}\frac{x}{\sqrt{\beta}}+1}{\sqrt[4]{2}\frac{x}{\sqrt{\beta}}-1}+2\tan^{-1}\left(1+\frac{\beta}{\sqrt{2}x^2}\right)\right] \tag{25}$$

This is the equation of the ray trajectory.

We can also use the expression obtained for dz and obtain all the parameters related to a waveguide.

a. **Ray Half-Period:**

This is the distance traveled along the axis of the guide between two successive turning points and is given by:

$$z_p = \int_{-x_{tp}}^{+x_{tp}} \frac{\beta dx}{\sqrt{n^2(x)-\beta^2}} \tag{26}$$

b. **Ray Path Length:**

This is the path length between any two turning points and the waveguide.

$$\frac{dz}{ds} = \cos\theta; ds = \frac{dz}{\cos\theta}$$

$$= \frac{\beta dx}{\cos\theta\sqrt{n^2(x)-\beta^2}} = \frac{n(x)dx}{\sqrt{n^2(x)-\beta^2}}$$

$$s = \int_{-x_{tp}}^{+x_{tp}} \frac{n(x)}{\sqrt{n^2(x)-\beta^2}} dx \tag{27}$$

c. **Optical Path Length:**

This distance traversed by the ray between two successive turning point and is given by:

$$dS = n(x)ds = \frac{n^2(x)dx}{\sqrt{n^2(x) - \beta^2}}$$

$$S = \int_{-x_{tp}}^{+x_{tp}} \frac{n^2(x)}{\sqrt{n^2(x) - \beta^2}} dx \tag{28}$$

d. Ray Transit Time:

This time is defined to be the time a ray of light takes to travel between two points in a waveguide.

$$t = \int_0^z \frac{1}{c} dS(z) \tag{29}$$

if z were exactly divisible by z_p, then

$$t = \frac{1}{c}\frac{z}{z_p} S' \tag{30}$$

where c is the speed of light.

Here, we have given only the outlines of a basic application of dynamic programming technique to the study of light propagation.

Discussion

Dynamic programming, a powerful optimization method was briefly described in this chapter. Calvo and Lakshminarayanan [6] have shown how this problem is equivalent to solving an eikonal equation (see also reference 7 and 8) and have illustrated the technique in the case of an ideal parabolic refractive index profile distribution.

Periodically segmented waveguides are used for improving conversion efficiency in frequency doubling in second harmonic generation with low power laser sources.

Using the dynamic programming method, a simple and easily implementable planar waveguide model and analysis has been presented [9,10].

This technique is based on the principle of optimality which is a statement about causality. Application of this technique results in a recurrence relation which can be very useful in solving the problem numerically. The technique involves dividing the problem into sub-problems, solving for the sub-problems and then using the functional relationship between the sub-problems to solve the whole problem. Since optimization is done at each level the number of algebraic operations involved is drastically reduced and hence is very useful in obtaining computational solutions.

Why are optimization techniques of interest? Currently optimization techniques based on the variational analyses related to the Sturm Liouville theory and the Rayleigh-Ritz method have been applied to waveguide design [e.g., 11, 12]. As we know, ray trajectories are solutions from the eikonal approach to the partial differential equation for the electromagnetic field propagating inside the medium. This technique has restrictions inherent in the integration condition of the ray path integral and on the required initial conditions. One of the main problems in optimizing the ray trajectory solution is the application of optimal interpolant procedures and associated complex computational methods [13, 14]. Dynamic programming which is related to the calculus of variations offers a simplified technique which could be applied to optimization problems in optical system design, minimization of aberrations in lens design, design of laser cavities, and so forth. There is much unexplored territory!

References

1. R. E. Bellman, *Dynamic Programming*, Pinceton Univ Press, Princeton NJ, (1957)
2. S.E Dreyfus, *Dynamic Programming and the calculus of variations*, Academic Press, NY, (1965)
3. S.E. Dreyfus and A.M Law, *The art and theory of dynamic programming*, Academic Press, NY, (1977)
4. R.E. Bellman, and S.E. Drefus, *Applied dynamic programming*, Princeton Univ Press, Princeton, NJ, (1962)
5. R.E. Bellman and R. Vasudevan, *Wave propagation,* Reidal. Derdrecht, Netherlands (1986)
6. M.L. Calvo, and V. Lakshminarayanan, *Light propagation in optical waveguides: a dynamic programming approach*, J. Opt. Soc. Am. A, **14**:872, (1997)

7. V. Lakshminarayanan, S. Varadarajam, ML Calvo, *A note on the applicability of dynamic programming to wave guide problems*, in Photonics 96: Proceedings of the international conference on photonics and fiber optics, ed. J.P. Raina, P.R. Vaya, Tata Mc Graw Hill, New Delhi, Vol. 1, Pp. 209-216, (1997)
8. V. Lakshminarayanan, S. Varadharajan, *Dynamic Programming, Fermat's principle and the Eikonal equation – revisited*, J. Optimization Theory and Applications, **95**, 713, (1997)
9. M.L. Calvo and V. Lakshminarayanan, *Optimal design using dynamic programming: Applications to gain guided segmented planar waveguides*, in Optics and Optoelectronics: Theory, Devices and Applications, Vol. 2, eds. O.P. Nijhawan, A. K. Gupta, A.K. Musla and K. Singh, Springer-Narosa, New Delhi, pages 1206-1214, (1999)
10. M.L. Calvo and V. Lakshminarayanan, *Spatial pulse characterization in periodically segments wave guides by using dynamic programming approach*. Optics Communications, **169**:223, (1991)
11. R.K. Laghu, and R.V. Ramaswamy, *A variational finite difference method for analyzing channel waveguides with arbitrary index profiles*, IEEE J. Quant. Electron. **QE-22**, 968, (1986)
12. A.Sharma, P. Bindal, *Variational analysis of diffused planar and channel waveguides and directional couplers*, J. Opt. Soc. Am. A, **11**, 2244, (1994)
13. B.D. Stone, G.W. Forbes, *Optical interpolants for Runge Kutta ray tracing in inhomogeneous media*, J. Opt. Soc. Am. A, **7**, 248, (1990)
14. J. Puchalsky, *Numerical determination of continuos ray tracing: The four component method*, Appl. Opt. **33**, 1900, (1994)

Appendix A: Geometric optics approximation and the eikonal equation

The basic concept in the geometric optics approximation in that in the optical range the propagation of waves can be represented (to a very high degree of accuracy) as the transfer of wave energy, which are described by geometric relations. Here we present a transition from the wave equation to an equivalent relation for rays. Consider the scalar wave equation for a light wave. This can be written as

$$\nabla^2\phi - \frac{1}{u^2}\frac{\partial^2\phi}{\partial t^2} = \nabla^2\phi - \frac{n^2}{c^2}\frac{\partial^2\phi}{\partial t^2} = 0 \qquad (1)$$

where we have introduced the refractive index n=c\v of the medium relative to vacuum. If n is a constant, for monochromatic waves we assume that

$$\phi = \psi(r)e^{-i\omega t}$$

where ω is the angular frequency. The above solution can be simplified by assuming that if ϕ is a plane wave,

$$\phi = \phi_o e^{i(k.r-\omega t)}$$

where k is the wave number ($= \frac{2\pi}{\lambda} = n\frac{\omega}{c}$)

If the direction of **k** is assumed to be along the z direction, then

$$\phi = \phi_{\circ} e^{ik_{\circ}(nz-ct)}$$

where $k_{\circ}$ is the wave number in vacuum ($k=nk_{\circ}$)

The plane wave solutions are not physically possible in an inhomogeneous medium because the variation in refractive index in the direction of propagation will bend the wave. We assume that the solutions are very nearly plane waves and write:

$$\phi = e^{A(r)+ik_{\circ}(S(x)-ct)} \tag{3}$$

where A and S are functions of positions. A is the amplitude of the wave and S would reduce to nz if n were a constant and can be recognized as being the optical pathlength. Both quantities are real. Taking the above solution, and substituting into the wave equation (1), we can write

$$ik_{\circ}[2\nabla A\cdot\nabla S+\nabla^2 S]\phi+[\nabla^2 A+(\nabla A)^2-k_{\circ}{}^2(\nabla S)^2+n^2k_{\circ}{}^2]\phi=0 \tag{4}$$

Since S and A are real measurable quantities the equation holds only if the two terms in the square brackets vanish:

$$\begin{aligned} &\nabla^2 A+(\nabla A)^2+k_{\circ}(n^2-(\nabla s)^2)=0\\ &\nabla^2+2\nabla A\cdot\nabla\cdot S=0 \end{aligned} \tag{5}$$

we make the assumption that n varies only slowly with distance, and is nearly a constant over distances of the order of wavelength. This implies that $\lambda < l$ where l is the dimension of any change in the medium. This is the fundamental geometric optics approximation.

The third term $k_{\circ}{}^2(n^2-(\nabla S)^2)$ is the most important term since $k_{\circ}{}^2 = 4\pi^2/\lambda_{\circ}{}^2$. This leads us to the fundamental equation of geometrical optics called the eikonal equation:

$$(\nabla S)^2 = n^2 \tag{6}$$

The gradient function S is directed along the normal to the surface S = constant. Therefore, the eikonal describes the constant-phase surfaces of a wave while grad A describes the propagation of light energy along a certain direction at a given point, and leads to the concept of a ray of light. The ray can formally be defined as a line the tangent to which at each point coincides with the vector grad S . The propagation of light can therefore be considered as the motion of light energy along the rays. The plane perpendicular to light rays (i.e., plane S = constant) is nothing but the wavefront. The geometric optics (ray) approximation is always valid when $\nabla^2 A / A$ is small in comparison to $1/\lambda^2$. That is, the wave amplitude varies significantly only at distances greater that λ. Physically, this implies that this term describes the bending of light by material objects-diffraction effects. Without going into detail, it can be stated that "classical mechanics corresponds to geometric optics limit of wave motion [1]" in which light rays orthogonal to wavefronts correspond to particle trajectories orthogonal to surfaces of constant S.

Reference:

1. H. Goldstein, Classical Mechanics, Addison Wesley, Cambridge, MA, (1956)

APPENDIX B: FERMAT'S PRINCIPLE FOR A GENERAL MEDIUM OF ARBITRARY ANISOTROPY

Upto this point, we have been dealing with Fermat's principle by assuming an isotropic, non-dispersive medium. Arley (Ref. 1) first showed that Fermat's principle is also applicable to the general case of an anisotropic medium in his studies of the Maxwell equations in the geometrical optical result. A simpler proof was given by Newcomb (Ref. 2) and is derived here.

Consider a simple monochromatic wave, with phase velocity

$$\phi = \omega[\psi(\mathrm{x}) - \mathrm{t}] \tag{1}$$

where ω is a constant and $\mathbf{X} = (x, y, z)$. We also define the vector

$$\mathrm{K} = (\kappa, \lambda, \mu) = \nabla\psi \tag{2}$$

with components

$$\kappa = \frac{k}{\omega} = \frac{\partial \psi}{\partial x}, \lambda = \frac{l}{\omega} = \frac{\partial \psi}{\partial y}, \mu = \frac{m}{\omega} = \frac{\partial \psi}{\partial z}$$

$$k = \frac{\partial \phi}{\partial x}, l = \frac{\partial \phi}{\partial y}, m = \frac{\partial \phi}{\partial z} \tag{3}$$

$$\omega = -\frac{\delta \phi}{\delta t}$$

and ψ satisfies the equation

$$F = (K, z) = 0 \tag{4}$$

where

$$F = \frac{1}{2}[a^2(\Gamma^2 + \lambda^2) + b^2\mu^2 - 1] \tag{5}$$

This is known as the "reduced eikonal" equation. In the above equation a and b are the principal phase velocities. In a spatially inhomogeneous medium in the z direction they are functions of z. We discard polarization in this analysis. In general ϕ will satisfy a first-order partial differential equation of the form

$$G(\omega, \mathbf{K}, \mathbf{X}) = 0 \tag{6}$$

The **X** dependence takes into account spatial inhomogeneities in the medium, and $\mathbf{K}(k, l, m) = \nabla\phi$. The reduced eikonal equation becomes

$$F(\omega, \mathbf{K}, \mathbf{X}) = 0 \tag{7}$$

or $F(\mathbf{K}, \mathbf{X}) = 0$ since we are dealing with a purely monochromatic wave.

It is well-known that the solution of a first order differential equation such as that given above can be reduced to an equivalent set of ordinary differential equations, the characteristic equations:

$$\mathbf{dx} : \mathbf{dk} = \mathrm{F}_{\mathbf{k}} : -\mathrm{F}_{\mathbf{X}} \tag{8}$$

where $F_{\mathbf{k}}$ and $F_{\mathbf{X}}$ or the partial derivatives of F or

$$\mathbf{X}'(\mathrm{s}) = \nu \mathrm{F}_{\mathbf{k}}(\mathbf{K}, \mathbf{X}) \tag{9}$$

$$\mathbf{K}'(\mathrm{s}) = -\nu \mathrm{F}_{\mathbf{X}}(\mathbf{K}, \mathbf{X}) \tag{10}$$

where ν is an arbitrary multiplier which depends on the choice of the parameter s.

The solution curves $\mathbf{X} = X(s)$ are the rays. We can also define the ray direction as the direction of signal propagation. Consider the interference between two neighboring solutions of the wave equation corresponding to the same value of ω. The two solutions are given by $\exp(i\omega\,\psi)$ and $\exp[(i\omega(\psi + \delta\psi)]$ where ψ satisfies $F(\mathbf{K}, \mathbf{X}) = 0$ and $\delta\psi$ at the same portion satisfies

$$\delta \mathrm{F} = 0 = \mathrm{F}_{\mathbf{k}} . \delta_{\mathbf{K}} = \mathrm{F}_{\mathbf{K}} . \nabla(\delta\psi) \tag{11}$$

The condition for constructive interference occurs at any part $\delta\psi = 0$ or alternately in terms of ray direction $d\mathbf{X}$

$$\mathrm{d}(\delta\psi) = \mathrm{d}\mathbf{X}.\nabla(\delta\psi) = \mathrm{d}\mathbf{X}.\delta\mathbf{K} = 0 \tag{12}$$

This above equation should be satisfied by any admissible $\delta\mathbf{K}$ which satisfies the expression for δF. This implies that the vectors dX and $F_{\mathbf{K}}$ are parallel. Next, consider the following (use is made of the symmetry of the tensor $\nabla\mathbf{K} = \nabla\nabla\psi$).

$$\nabla \mathrm{F} = \mathrm{F}_{\mathbf{X}} + \nabla \mathbf{K} \mathrm{F}_{\mathbf{K}} = 0 \tag{13}$$

$$K' = X'.\nabla K = \nu F_K.\nabla K = \nu \nabla K F_K = -\nu F_X \tag{14}$$

This is nothing but a generalized form of the law governing refraction of waves by spatial inhomogeneities in the medium.

Consider the case of a uniaxial crystal, with optic axis in the *z*-direction. If, as before, *a* and *b* are the principal phase velocities and ϕ the wave phase, the eikonal equation for the *E* wave is

$$a^2(k^2 + l^2) + b^2 m^2 = \omega^2 \tag{15}$$

where $\omega = \partial\phi / \partial t$.

The characteristic equations are:

$$x' = \mu a^2 \kappa, y' = \mu a^2 \lambda, z' = \mu b^2 \mu$$
$$K' = \lambda' = 0, \mu' = -\mu a a_z(\kappa^2 + \lambda^2) - \mu b b_z \mu^2$$

Eliminating κ, λ and μ and putting $F = 0$, we get

$$(x'/\mu a^2)' = (y'/\mu a^2)' = 0 \tag{16}$$

$$\mu(z'/\mu b^2)' - a_2(x'^2 + y'^2)/a^3 + b_z z'^2/b^3 = 0 \tag{17}$$

where $\mu = [(x'^2 + y'^2)/a^2 + z'^2/b^2]^{1/2}$. These are the differential equations which describe the rays.

If we make a suitable choice of the parameter *s* such that

$$b^2(x'^2 + y'^2) + a^2 z'^2 = b^2/a^2$$
$$\Rightarrow \mu a^2 = 1$$

The ray equations reduce to

$$x'' = y'' = 0 \tag{18}$$

$$(a^2z'/b^2)+a_z(x'^2+y'^2)/a+a^2b_zZ'^2/b^3=0 \tag{19}$$

The ray direction is in the direction of the Poynting vector.

The wavefront is defined as a moving contour surface of constant phase ϕ. The moving wavefront successively coincides with each of the stationary surfaces of constant $\psi(x)$. This means

$$d\psi = dt \; d\psi = dt \tag{20}$$

For the change in Ψ in the time *dt* following the wavefront. This can also be written as:

$$dt = d\mathbf{x}.\nabla\psi = \mathbf{K}.\mathbf{X}'ds \tag{21}$$

Let A_0 and A_1 be any two fixed points on a particular ray *R*, to which we arbitrarily give the parameter values $S = 0$ and $S = 1$, and *P* denote the intersection of the moving wavefront with *R* the ray.

The time required for passage of *P* from A_0 and A_1 is given by

$$T = \int_0^1 \mathbf{K}.\mathbf{X}'(s)\,ds \tag{22}$$

It should be noted that *R* is the path of stationary *T*. *R* is determined as the path of minimum *T* between the given end points. This is nothing but Fermat's principle. Note that the "time of passage" in the Fermat prinbciple is specifically that of a point on the ray carrying a constant value of the phase ϕ. As a result of dispersion, it will in general differ from the measured travel time of a wave signal from A_0 and A_1 (Ref. 3).

By solving the system of equations

$$F = 0.X' = \nu F_{\mathbf{k}} \tag{23}$$

We can represent **k** and ν as functions of **X** and **X′**, as

$$\mathbf{K.X}' = L(\mathbf{X}, \mathbf{X}') \tag{24}$$

Therefore, the integrand of T is expressible as a function of **X** and **X′**.

Fermat's principle requires us to compute the integral not only along the ray, but also along various neighboring paths which are not themselves rays. How can this be done? The value of **K** to be used at a particular point is that which is appropriate to the ray having the same tangent vector, at the point in question as the given path. If we do a total differentiation of $F = 0$.

$$dF = 0 = F_{\mathbf{k}}.d\mathbf{k} + F_x.dx \tag{25}$$

$$\begin{aligned} dL &= L_x.d\mathbf{x} = L_x.dx' = d(\mathbf{k}.x') = \mathbf{K}.dx' + X'.d\mathbf{K} \\ &= \mathbf{k}.dx' + \nu F_{\mathbf{k}}.d\mathbf{K} = \mathbf{K}.dx' - \nu F_x.d\mathbf{x} \\ &\Rightarrow L_x = -\nu F_x, \quad L_{x'} = \mathbf{K} \end{aligned} \tag{26}$$

The second order equation for $\mathbf{x}(s)$ to which $\mathbf{K}'(s) = -\nu F_x(\mathbf{K}, \mathbf{X})$ reduces in general:

$$(L_{x'})' - L_x = 0 \tag{27}$$

This is nothing but the Euler-Lagrange condition for stationary T'.

In the example of the uniaxial crystal given above,

$$L(X,X') = [(x'^2 + y'^2)/a^2(z) + z^2/b^2(z)]^{1/2} \tag{28}$$

A differential geometric analysis of light transmission in an anisotropic medium.

For the anisotropic medium, Shen et al [4] have generalized Fermat's principle and the optical metric and have developed a Riemannian geometrical description of the space geometry. For an anisotropic medium, the dielectric constant becomes a symmetrical tensor in a three-dimensional space. Shen et al have established the relationship between the metric coefficients of a three dimensional Riemannian manifold and the dielectric tensor. There is a physical connection between the three dimensional geometry corresponding to geometrical optics and four-dimensional space-time geometry corresponding to metric optics. The geodesic equations in such a space determines the light track and geodesic deviation equations can describe focussing and defocusing of an optical beam and optical transmission. (see references 5- 7).

Consider a static, nonconducting and non magnetic anisotropic medium. It is known that the dielectric tensor ε is symmetric and has only 6 instead of nine components. The dielectric tensor reduces to a diagonal form

$$\varepsilon = \begin{bmatrix} \varepsilon_{11} & \varepsilon_{12} & \varepsilon_{13} \\ \varepsilon_{21} & \varepsilon_{22} & \varepsilon_{23} \\ \varepsilon_{31} & \varepsilon_{32} & \dot{\varepsilon}_{33} \end{bmatrix} = \begin{bmatrix} n_1^2 & & \\ & n_2^2 & \\ & & n_3^2 \end{bmatrix} \tag{29}$$

if the coordinate system of principal dielectric axes is used. On the other hand for an inhomogeneous anisotropic medium, $\varepsilon_{ij} = \varepsilon_{ij}(x, \dot{x})$ and the six components ε_{ij} of the dielectric tensor will vary with x and $\dot{x}$. In this case both values of the principal dielectric constants and the directions of principal axis will vary. Shen et al made the basic assumption that the anisotropy is globally aligned and $\varepsilon_{ij} - \varepsilon_{ij}(x)$. This results in a diagonal form of the dielectric tensor

Fermat's principle

$$\delta \int nds = \delta \int ndl = 0 \tag{30}$$

can be written in this case as

$$\delta \int nds = \delta \int \left(n_1{}^2 dx^2 + n_2{}^2 dy^2 + n_3{}^2 dz^2\right)^{1\backslash 2} \tag{31}$$

The element ds characterizes the geometric structure of a three dimensional manifold associated with the anisotropic medium and

$$\begin{aligned} ds^2 &= n_1{}^2 dx^2 + n_2{}^2 dy^2 + n_3{}^2 dz^2 \\ &= g_{ij}(x) dx^i dx^j \end{aligned} \tag{32}$$

where $g_{ij}(x)$ is the metric tensor of the manifold

$$\begin{aligned} g^{ij} = n_i^2 \delta_{ij} &= \begin{cases} n_i{}^2 & (i = j) \\ 0 & (i \neq j) \end{cases} \\ & \qquad i, j = 1,2,3 \\ &= \begin{bmatrix} n_1{}^2 & & \\ & n_2{}^2 & \\ & & n_3{}^2 \end{bmatrix} \end{aligned} \tag{33}$$

The corresponding inverse metric tensor has the form

$$g_{ij} = \begin{cases} n_i^{-2} & i = j \\ 0 & i \neq j \end{cases} \quad i, j = 1,2,3$$

$$= \begin{bmatrix} \frac{1}{n_1^2} & & \\ & \frac{1}{n_2^2} & \\ & & \frac{1}{n_3^2} \end{bmatrix} \tag{34}$$

Since the metric has been defined, we can derive the geodesic equation of the light track in the static anisotropic medium using standard methods of differential geometry:

$$\frac{dx^i}{ds^2} + \Gamma^i_{jk} \frac{dx_i}{ds} \frac{ds^k}{ds} = 0 \tag{35}$$

where Γ^i_{jk} is the affine connection coefficient and is given by

$$\Gamma^i_{jk} = \frac{1}{2} g^{il} \left(\frac{\delta g_{lj}}{\delta x^k} + \frac{\delta g_{lk}}{\delta x^j} - \frac{\delta g_{jk}}{\delta x^l} \right) \tag{36}$$

substituting for g_{ij} etc. we get the connection coefficient as

$$\Gamma^i_{jk} = \Gamma^i_{kj} = \begin{cases} n_i^{-1} n_{i,k} & i = j \\ -n_i^{-2} n_j n_{j,i} & i \neq j = k \\ 0 & i \neq j \neq k \neq i \end{cases} \tag{37}$$

where $n_{ij} = \frac{\delta n_i}{\delta x^j}$

Using the fact $(x^1 x^2 x^3) = (x\ y\ z)$ and substituting into the equation for the geodesics, after much algebraic manipulation we get

$$\frac{d}{ds}\left\{(i \quad j \quad k)\begin{bmatrix} n_1^2 & & \\ & n_2^2 & \\ & & n_3^2 \end{bmatrix}\begin{bmatrix} dx/ds \\ dy/ds \\ dz/ds \end{bmatrix}\right\}$$

$$= \frac{1}{2}\begin{pmatrix} \frac{dx}{ds} & \frac{dy}{ds} & \frac{dz}{ds} \end{pmatrix}\nabla\begin{bmatrix} n_1^2 & & \\ & n_2^2 & \\ & & n_3^2 \end{bmatrix}\begin{bmatrix} dx/ds \\ dy/ds \\ dz/ds \end{bmatrix} \tag{38}$$

or in matrix form

$$\frac{d}{ds}\left[eN^2\left(\frac{dx}{ds}\right)\right] = \frac{1}{2}\frac{dx}{ds}\nabla N^2\left(\frac{dx}{ds}\right)^T \tag{39}$$

where $\mathbf{e}\mathbf{N}^2 = (ijk)$ are unit vectors in the directions of the three coordinate axis.

$\left(\frac{dx}{ds}\right)^T$ is nothing but the transpose of the row matrix $\frac{dx}{ds} = \left[\left(dx/ds\right)\left(dy/ds\right)\left(dz/ds\right)\right]$

This equation is simply the generalized eikonal equation that governs light track in a static anisotropic medium. If n_i=n, the equation reduces to the familiar eikonal equation (Appendix A).

If we calculate the curvature tensor R^i_{jkl} and the curvature R, we can determine the geometrical structure of the manifold and the behavior of light rays in the anisotropic medium. The curvature tensor is defined to be

$$R^i_{jkl} = \Gamma^i_{jl,k} - \Gamma^i_{jk,l} - \Gamma^m_{jl}\Gamma^i_{mk} - \Gamma^m_{jk}\Gamma^i_{ml} \tag{40}$$

where $\Gamma^i_{jk,l}$ is the first order ordinary partial derivative of the affine connection Γ^i_{jk} with respect to the co-ordinate x^l. Differentiating the expression for Γ^i_{jk} gives

$$\Gamma^{i}_{jk,l} = \Gamma^{i}_{kj,l} = \begin{cases} n_i^{-1} n_{i,kl} - n_i^{-2} n_{i,k} n_{i,l} & (i = j) \\ 2n_i^{-2} n_j n_{j,i} n_{i,l} - n_i^{-2}\left(n_{j,l} ni + n_j n_{j,il}\right) & (i \neq j = k) \\ 0 & (i \neq j \neq k \neq i) \end{cases} \tag{41}$$

where $n_{i,jk} = \delta^2 n_i \big/ \delta x^i \delta x^k$

substituting these expressions into the equation for the curvature tensor, we get:

$$R^{i}_{jkl} = -R^{i}_{jlk} = \begin{cases} \frac{1}{4} n_i^{-2} \left\{ \begin{array}{l} 2\left(n^2_{i,jj} + n^2_{j,ii}\right) \\ + n_p^{-2} n^2_{i,p} n^2_{j,p} - n_i^{-2}\left[\left(n^2_{i,j}\right)^2 + n^2_{i,i} n^2_{j,i}\right] \\ - n_j^{-2}\left[\left(n^2_{j,c}\right)^2 - n^2_{j,j} n^2_{i,i}\right] \end{array} \right\} & (i \neq j, i = l, j = k) \\ \frac{1}{4} n_i^{-2} \left\{ \begin{array}{l} 2n^2_{i,jk} - \\ \left(\begin{array}{l} n_i^{-2} n^2_{i,j} n^2_{i,k} + n_j^{-2} n^2_{i,j} n^2_{j,k} \\ + n_k^{-2} n^2_{i,k} n^2_{k,j} \end{array} \right) \end{array} \right\} & (i \neq j \neq k \neq i, i = l) \\ \frac{1}{4} n_i^{-2} \left\{ \begin{array}{l} 2n^2_{j,ik} - \\ \left(\begin{array}{l} n_i^{-2} n^2_{j,i} n^2_{i,k} + n_j^{-2} n^2_{j,i} n^2_{j,k} \\ + n_k^{-2} n^2_{j,k} n^2_{ki} \end{array} \right) \end{array} \right\} & (i \neq j \neq k \neq i, j = l) \end{cases} \tag{42}$$

and the remaining components are zero.

The curvature scalar R defined by

$$R = R^{i}_{i} = g^{ik} R_{ik} = g^{ik} R^{l}_{ikl} \tag{43}$$

which is simply

$$R = \frac{1}{4} n_i^{-2} n_j^{-2} \left\{ 2\left(n_{i,jj}^2 + n_{j,ii}^2\right) - 2n_i^{-2}\left[n_{i,i}^2 n_{j,i}^2 + (n_{i,j}^2)^2\right] + n_k^{-2} n_{i,k}^2 n_{j,k}^2 \right\} \tag{44}$$

where the summation is done over the dummy indices $i,j,k = 1,2,3$ and $i \neq j \neq k \neq i$

The curvature tensor and curvature scalar are completely determined by the squared refractive index of the anisotropic medium. These quantities completely determine the geometric structure of the manifold and hence the behavior of light rays in the medium. The null geodesic equation governs the light track and the geodesic deviation equation will describe the focusing and defocusing of light rays. The curvature tensor and curvature can also describe birefractive phenomena in the anisotropic medium. The reader is referred to the literature for details.

References

1. N. Arley, K.I. Danslee, V. Denskeb, Selskab, Mat-Fys, 22: 1, 1945, M.Kline and I.W. Kay, *Electromagnetic theory and Geometrical optics*, Wiley-interscience, New York, 1965, see also Yu A. Kravtsov and Yu. I. Ovlov, *Geometrical optics of inhomogeneous media*, Sprenger-verlag, Berlin, (1990).
2. W.A. Newcomb, *Generalized Fermat principle*, Am. J. Phys., **51**: 338, (1983)
3. F.A. Jenkins and H.E. White, *Fundamentals of optics*, 4th edition, McGraw-Hill, NY (1976), Sec. 12.7.
4. W. Shen, J.Zhang, S.Wang, S.Zhu, *Fermat's principle, the general eikonal equation and space geometry in a static anisotropic medium*, J. Opt. Soc. Am. A, **14**:2850, (1997)
5. W.Shen, S.Zhu, X. Deng, *The light tracks in the optical fibers with 2 types of parabolic refractive indices*. Chin. J. Lasers B, **5**,516, (1996)
6. H.Guo and X.Deng, *Differential geometric methods in the study of optical transmission (scalar theory) I. Static transmission case*, J. Opt. Soc. Am. A, **12**, 600, (1995);
7. H.Guo and X.Deng, *Differential geometric methods in the study of optical transmission(scalar theory) II, time dependent transmission case*, J. Opt. Soc. Am. A, **12**, 607, (1995)

APPENDIX C: RAY PROPAGATION AND SYMPLECTIC TRANSFORMATIONS

As usual in Geometric optic consider two planes, the input and output planes, and let the z axis be the optic axis. The optical z axis is the direction of light propagation. Light enters the optical system from the left at the input plane and exits at the right output plane. Following the notation of chapter 1, let (q_{x_i}, q_{y_i}) be the point of intersection of ray with input plane and corresponding optical direction cosines (p_{x_i}, p_{y_i}) be

$$\begin{aligned} p_{x_i} &= n_i \sin\alpha_{x_i} \\ p_{y_i} &= n_i \sin\alpha_{y_i} \end{aligned} \tag{1}$$

where n_i is given by the local value of the refractive index, $n_i \equiv n(q_{x_i}, q_{y_i}, z)$ and $(\alpha_{x_i}, \alpha_{y_i})$ are the angles the ray makes with the q_x-z and q_y-z planes. Similar notation is used for the out-going ray at the output plane.

The fundamental problem in geometrical optics is to determine the input-output relations of the optical system.

$$\begin{aligned} q_\circ &= q_\circ(q_i, p_i) \\ p_\circ &= p_\circ(q_i, p_i) \end{aligned} \tag{2}$$

These input-output relations corresponds to a trajectory in phase space from the point $p_1 \equiv (q_{x_i}, q_{y_i}, p_{x_i}, p_{y_i})$ at z_i to $p_\circ \equiv (q_{x_\circ}, q_{y_\circ}, p_{x_\circ}, p_{y_\circ})$ at z_f and these trajectories are governed by the Hamiltonian and the transformation is a canonical/symplectic transformation and is a straight forward consequence of the structure of the equations of motion.

In chapter 8, we have introduced the fact that this symplectic transformation is a mapping given by the matrix M. Consider the 4 x 4 antisymmetric matrix J (equation 8.8)

$$J = \begin{pmatrix} 0 & I \\ -I & 0 \end{pmatrix} \tag{3}$$

Each entry in the J is a 2x2 matrix.

$$I = \begin{pmatrix} 1 & 0 \\ 0 & 1 \end{pmatrix}$$
$$O = \begin{pmatrix} 0 & 0 \\ 0 & 0 \end{pmatrix} \tag{4}$$

The matrix J has the properties

$$J^{-1} = -J, \qquad J^2 = -I$$
$$J^T = -J, \qquad \det J = 1 \tag{5}$$

where J^T denotes the transpose. Let the components of the phase space vector be indexed as 1,2,3,4, such that

$$u^T = (q_x, q_y, p_x, p_y)$$
$$\equiv (u_1, u_2, u_3, u_4) \tag{6}$$

The equations of motion

$$\frac{dq_x}{dz} = \frac{dH}{dp_x} \qquad \frac{dq_y}{dz} = \frac{dH}{dp_y}$$
$$\frac{dp_x}{dz} = \frac{dH}{dq_x} \qquad \frac{dp_y}{dz} = \frac{dH}{dq_y} \tag{7}$$

can be written succinctly as

$$\frac{du}{dz} = J_{jl} \frac{\delta H}{\delta u_l} \qquad j,l = 1,2,3,4 \tag{8}$$

Now, if we transform the vector u to v

$$v_j = v_j(u_1, u_2, u_3, u_4) \tag{9}$$

v obeys the equation

$$\frac{dv_j}{dz} = M_{jl} \cdot J_{lk} \cdot M_{hk} \frac{\delta H}{\delta V_k} \tag{10}$$

where M_{ij} denotes the entries of the Jacobian matrix associated with the transformation above

$$M_{ij} = \frac{\delta v_i}{\delta_{u_j}} \tag{11}$$

If we require that the transformation leave the Hamiltonian invariant, it implies the matrix relation

$$MJM^T = J \tag{12}$$

This is the condition that M be a symplectic matrix. It is an easy matter to verify that the 2m x 2m non-singular matrices obeying the above relation forms a group under matrix multiplication. The associativity, identity and unique inverse are readily shown while the closure property also follows from the relationship (i.e., if M_1 and M_2 obey $MJM^T = I$) so will $(M_1M_2)J(M_1M_2)T = I$. In mathematical terms we say that M belongs to the symplectic group Sp(4,R). It is also very easy to show that M^{-1} also obeys the above equation. Also,

$$M^T JM = I \tag{13}$$

other properties include

$$\det M = \pm 1 \tag{14}$$

The sign of the determinant can be fixed by noting that in a phase space transformation elements of volume transform through the determinant of the Jacobian matrix relative to the transformation according to $dV_i = \det M dV_o$. Since by Liouville's theorem of phase space volume invariance, $dV_o = dV_i$ which puts the condition that $\det M = \pm 1$.

Also if the transformation between the vectors V and u is linear the matrix elements if M are just the coefficients of the linear combination

$$v_i = \sum_{j=1}^{4} M_{ij} u_j \tag{15}$$

Because of the symplecticity relation (12) or (13) the 16 entries of the 4x4 matrix M are not all independent. Only 10 can be arbitrarily chosen, the others being fixed. This comes from the general property of 2mx2m symplectic matrices for which only m(2m+1) elements can be arbitrarily assigned. The rest are fixed by the equation (12) or (13).

INDEX

R

MIX
Papier aus verantwortungsvollen Quellen
Paper from responsible sources
FSC® C105338

If you have any concerns about our products,
you can contact us on
ProductSafety@springernature.com

In case Publisher is established outside the EU,
the EU authorized representative is:
Springer Nature Customer Service Center GmbH
Europaplatz 3, 69115 Heidelberg, Germany

Printed by Libri Plureos GmbH
in Hamburg, Germany